Designing Work for Sustainability

Naresh C. Singh
Carol Amaratunga
Cynthia Pollock Shea
Jacqueline Romanow

IISD INTERNATIONAL INSTITUTE FOR SUSTAINABLE DEVELOPMENT
INSTITUT INTERNATIONAL DU DÉVELOPPEMENT DURABLE IIDD

TABLE OF CONTENTS

PREFACE

Progress in sustainable development is made when there are mutually reinforcing advances in the social, economic and ecological spheres of human interactions with nature. Progress in any one sphere without consideration of its impacts on the others could be self-defeating. In recognition of these interlinkages, IISD offered to the First PrepCom of the World Summit for Social Development (WSSD) a report on *Sustainable Development and the World Summit for Social Development: conceptual and practical linkages among sustainable development, poverty eradication, productive employment and social integration.*

The underlying principle of our contributions to the WSSD is the importance of fostering sustainable livelihoods.

In the present document, we make a further contribution to the WSSD process by focusing on the issue of employment in relationship to sustainable development strategies and thus to social development and poverty eradication. While this report directs several proposals to the Summit and its preparatory process, the substantive issues are discussed in enough detail to be useful to interested persons outside of the Summit process.

IISD plans to make a third contribution to the Summit process, which will focus on methodological issues in galvanising the transition from poverty to sustainable livelihoods.

Arthur J. Hanson
President and C.E.O.
IISD

ACKNOWLEDGMENTS

Much of the information for the examples and case studies on employment for sustainable development are drawn from materials prepared for the IISD meeting on "Employment and Sustainable Development" held in Winnipeg, June 24th-26th, 1994, at the request of the Minister of Human Resources of the Government of Canada. The individual sources of the cases are acknowledged in the text. We are grateful to our colleagues, Stephan Barg and Rob Kerr who organised the meeting and made the materials available to us.

Naresh Singh
Carol Amaratunga
Cynthia Pollock Shea
Jacqueline Romanow

SUMMARY

The fact that poverty is a serious and growing problem in a world with abundant resources is proof that the global economy has failed to meet its principle objective – namely to mete out adequate goods and services to the earth's population. Mass unemployment in both the industrialized and non-industrialized world is leading to accelerating poverty, misery and social decay. 120 million people are officially registered as unemployed, while unofficially, the figure may be as high as 820 million (nearly 30% of the total labor force). Concurrently, global environmental pollution and ecological degradation, both products of short-sighted traditional economic development, are constraining the earth's capacity to produce the resources required to maintain its growing population. Unfortunately, current macro trends, including global re-structuring and expanding free trade, hold little promise for an improved future. What is needed is significant change.

Fundamentally, the global economy must be structured in such a way that it works with, not against, the natural environment. Wholesale resource depletion is not sustainable. Neither is an economic system which treats humanity as its slave. A healthy, sustainable global economy would provide meaningful livelihoods which do not destroy the natural environment or the human psyche. At the same time, such a system cannot be premised on unrealistic, utopian-style, dreams. For example, it is not realistic to assume that human beings will not try to maximize their economic well-being. Effort is needed, therefore, to ensure that the linkages between the well-being of the individual, society, the ecosystem and the natural environment are made explicit.

The dichotomy between "core" and "margin" is misleading in that it detracts from the viability and value of cultures unlike industrial capitalism. The whole concept of economic development must be redefined to incorporate local culture and values. The fulfillment of human needs and the provision for the development of human potential must become the key focus of economic development. At the same time, there is an immediate need to recognize and address more immediate and practical concerns.

The case studies offered in the document provide excellent examples of the immediate potential which exists within the realm of sustainable economic development in terms of jobs. Practical opportunites currently exist within the following areas:

- Ecosystem based forest management, which accounts for a holistic relationship between ecosystems and cultural values, offers a viable and sustainable alternative to current forestry practices.
- Significant employment opportunities exist in evaluating building performance, advising building owners on cost-effective upgrades, manufacturing, marketing and selling energy efficient products and installing improved technologies.
- Energy conservation represents a new and dynamic sector of the economy which is capable of creating thousands of new job opportunities.
- Through resource restoration and sustainable resource harvesting, aboriginal peoples can regain control and management of traditional territories.
- Small community-based organizations are best able to assess the capabilities and needs of the community and, as such, directly facilitate local development efforts.
- Aquaculture development has the potential to provide both much needed foodstuffs and employment at a time when natural foodstocks are in serious decline.
- Through a holistic design process, retrofitting technologies can provide jobs, ensure sustainability and educate the public.

The core issues of the WSSD viz employment, poverty and social integration are key concerns for sustainable development. Consistency and continuity in action and decision making in the international community require the preparations for the Social Summit to take full account of the UNCED process and indeed to advance much needed progress in that regard.

This report is structured to contribute to the declaration, the programme of action and any other outcome of the WSSD. It concludes by offering some concrete suggestions for that outcome, including, for example, the creation of an International Council for Social Progress.

CHAPTER 1 INTRODUCTION

This report examines the concept of sustainable employment and its implications for poverty eradication and social integration as envisaged by the preparatory documents of the World Summit for Social Development (WSSD). While the primary readership is intended to be those active in the Summit and its preparatory process; the concepts, principles, and proposals and their inter-relationships would be of interest to the broader sustainable development readership.

In preparing this document, we have considered the note by the UN Secretary-General entitled "Outcome of the World Summit for Social Development: Draft Declaration and Draft Programme of Action (a/CONF.160/PC/L.13 dated 3rd June, 1994) as well as documents of related UN conferences such as the International Conference on Population and Development (1994) and UNCED (1992). Our intention is to build on the ideas contained in these and other sources as well as to offer alternative view points where we feel necessary.

While this report is neither a critique nor a response to the note by the Secretary-General (L.13); the parallels between the two documents are deliberate and intended to facilitate the use of this report during the preparatory deliberations leading to the Summit, while maintaining a logical flow for the more general readership.

While substantively different, the contents of the Draft Declaration parallels our sections on the Problem and the Vision. The section in the Draft Programme of Action on An Enabling Environment considers some of the issues we discuss in the subsection on Fundamental Transitions.

Using sustainable development as the major theme we then discuss in Chapters 3 and 4 issues pertinent to the remainder of the action plan. This includes issues of poverty eradication and social integration as they result from sustainable employment and the enhancement of opportunities for each human being to make a living in a decent and sustainable manner. This approach allows us to offer substantive information and case studies on the issue of sustainable employment, an area in which there is a relative global shortage of concrete ideas.

The final chapter calls for concrete and tangible summit outputs and commitments and includes suggestions on arrangements for implementing whatever action plans are agreed on.

It is our hope that this contribution will provide ideas and insights which will strengthen the process and outcomes of the summit and thus promote and accelerate the global transition to sustainable social development.

Since several of the concepts basic to this discourse are gathering increasing loads of baggage and confusion, we state below how we use these terms and how they are related to each other.

CONCEPTS AND LINKAGES

Sustainable Development

The concept of sustainable development implies a process of change in which the exploitation of resources, the direction of investments, the orientation of technological development and exchange, and institutional change reflect both future and present needs. The notion of equity is seen as central to sustainable development and implies dramatic re-distribution of assets and the enhancement of capabilities and opportunities of the disadvantaged.

At the practical and operational level sustainable development means ensuring self-sustaining improvements in productivity and quality of life of communities and societies including access to basic needs such as education, health, nutrition, shelter and sanitation; as well as employment and food self-sufficiency; ensuring that production processes do not overexploit the carrying and productive capacities of the natural resource base and compromise the quality of the environment, thus limiting the options of the poor, the present and future generations; and ensuring that people have basic human rights and freedoms to participate in the political, economic, social and environmental spheres of their communities and societies.

Whereas previous attempts at so called "economic development" were typically in opposition to the natural environment, sustainable development aims to work with the natural order to ensure the provision of requisite resources for human survival and growth in both the short and long run.

Sustainable Employment/Sustainable Livelihoods

If we wish to look at the concepts of jobs in terms of sustainability it is important to look beyond the conventional notion of employment. Full-time permanent jobs are, due to changes in the global workplace, becoming a thing of the past. People can no longer depend on one company to provide them with a job for life. Instead a broader strategy must be utilised. This concept is best understood through the notion of sustainable livelihoods. Chambers and Conway (1992) propose a working definition of sustainable livelihoods as follows:

> "A livelihood comprises the capabilities, assets (stores, resources, claims, and access) and activities required for a means of living; a livelihood is sustainable which can cope and recover from stress and shocks, maintain and enhance the capabilities and assets and provide sustainable livelihood opportunities for the next generation; and which contributes net benefits to other livelihoods at the local and global levels and in the short and long term."

Alternatively we might define sustainable livelihoods as concerned with people's capacities to generate and maintain their means of living, enhance their well-being and that of future generations. These capacities are contingent upon the availability and accessibility of options which are ecological, socio-cultural, economic and political and are predicated on equity, ownership of resources and participatory decision making. Both the notions of sustainable development and sustainable livelihoods incorporate the idea of change and uncertainty. (Singh and Titi, 1994)

Poverty

A number of definitions of poverty have been advanced and a standard vocabulary has emerged such as poverty line, absolute and relative poverty, pockets, and mass poverty. These definitions have been arrived at through two processes outlined by Amartya Sen (1981), namely, identification of the poor through the specification of a set of basic or minimum needs and the inability to meet those needs; and aggregating the characteristics of the set of poor people into an overall image of poverty.

- **Poverty line:** the level of minimum household consumption that is socially acceptable – what Rowntree (cited in Singh and Titi, 1994) calls "primary poverty". It is usually calculated on the basis of an income of which roughly two-thirds would be spent on a "food basket" which provides the least-cost essential calories and proteins.
- **Absolute and relative poverty:** the phenomenon of absolute poverty is confined mostly to developing countries where the absolute poor are those who fall below the minimum standard of consumption (poverty line). Relative poverty, on the other hand, exists above the poverty line and is perceived as a state of deprivation relative to existing societal norms of income and access to social amenities, and is a concept more frequently used as describing poverty in modern industrialised economies.

- **Pockets and mass poverty:** the phase, pockets of poverty, has been applied to localized poor communities in the midst of affluence in the developed countries where the assumption is that the problem is relatively insignificant and can be easily dealt with. On the other hand, mass poverty has been used to describe poverty in the developing countries where the poor constitute a major fraction of the population and where it is increasingly becoming difficult to conceptually isolate the poor. (Singh and Strickland, 1993)

Robert Chambers (1983) maintains that the poor are characterized by: isolation, due to their peripheral location away from the centres of trading, discussions and information, lack of advice from extension workers in agriculture, forestry and health; and in most cases lack of means for travel. The poor remain powerless as long as they are isolated and vulnerable, because they have few buffers against contingencies. Small needs are met by drawing on meagre reserves of cash, by reduced consumption, by barter or by loans from friends, relatives and traders; and powerlessness, as a result of their ignorance of the law, lack of access to legal advice, competition for employment and services with other in similar conditions.

For our purposes, we adopt a definition of poverty as a condition of lack of access to options and entitlements which are social, political, economic, cultural and environmental. To the concept of poverty we link the notion of impoverishment. Impoverishment is an active process that leads to diminished access to options and entitlements. Poverty is the product of an inadequate livelihood base. (Singh and Titi, 1994). Poverty eradication can thus best be facilitated through the procurement of sustainable livelihoods for the earth's population. Impoverishment is continuously reproduced and generated by a number of currently active global mechanisms including environmental degradation, resource depletion, inflation, unemployment and debt. These mechanisms have set in motion the erosion of safety nets and the widening gap between rich and poor nations. In developing countries, processes of impoverishment arising from patterns of colonization and commodity exploitation were exacerbated by subsequent post-colonial attempts at addressing poverty through modernization approaches which fostered modes of industrialization and trade dependent upon imported technology and capital. Such trends contributed to increased stress on natural industrialization and trade dependent upon imported technology and resources and patterns of production and consumption which were incompatible with the long-term requirements of sustainable development. These have included rising landlessness, alienation from productive resources, and increased migration to urban areas already under stress from economic and environmental pressures.

Empowerment for sustainable development

This means access to options such as cultural and spiritual space, recognition and validation of endogenous knowledge, entitlements to land and other resources, income, credit, information, training and participation in decision-making to meet today's needs without foreclosing future options.

Empowerment

Empowerment reaffirms the principles of inclusiveness (i.e. engaging the relevant stakeholders in a process), transparency (i.e., openness) and accountability (i.e., giving legitimacy to any process and decisions reached), which are held in common with notions of democracy and sustainable development as articulated at the 1992 Earth Summit and have come to be known collectively as "the Rio way".

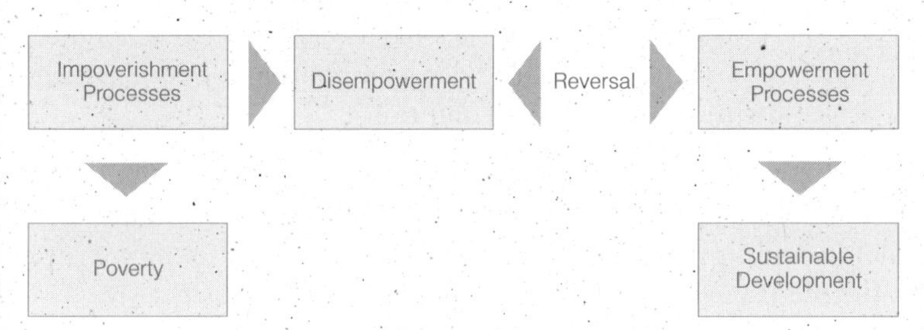

Interrelationship Processes of Empowerment, Disempowerment and Impoverishment and Sustainable Development

Social Cohesion and Collective Security

The 1994 Human development report warns "The world can never be at peace unless people have security in their daily lives." In an increasingly interconnected global village, phenomena like famines, pollution, drug trafficking, social disintegration, ethnic conflict and terrorism are no longer isolated events (Human Development Report, 1994). Consequences are felt around the earth. As such, human security must be approached at a world-wide level.

The most crucial aspect of human security is that which is found in the context of day to day life. When people do not have access to an adequate resource base from which to secure their livelihood, the resultant instability can lead to familial unrest, violence, criminal activity and social breakdown. With access to sustainable, secure livelihoods, human beings are better able to deal with the stress and changes embodied in sudden external shocks through coping and adaptive strategies. It is only through such security that world peace and social harmony can be fostered. With sustainable livelihoods, humans are able to pursue their own personal development as human beings. The end product is a healthier, more resilient social fabric.

Social Development

As noted in the Draft Program of Action for the second session of the Preparatory Committee the fundamental goal of the World Summit for Social Development is to "promote social progress and better standards of life in larger freedom." This provides a concise yet broad understanding of what is meant by the term "social development." Through social development, human beings are freed from the burdens of immediate survival and able to pursue broader intellectual and personal interests. Human beings are inherently social and creative beings and thus are only able to reach the fullest of human potential when these faculties are permitted to be explored and developed. Social development is therefore an essential ingredient to social cohesion. Social development allows for the fulfilment of individual potential which serves as a necessary foundation for a society which is fundamentally integrated, secure and healthy.

Implications for the Global Economy

Many people define themselves and their role in society in terms of their jobs. They want to feel proud of the work they perform and the way in which they perform it. In addition to receiving a paycheck, they want the intangible satisfaction of knowing that their skills and knowledge are being fully tapped and put to good use.

From within the paradigm of sustainability, a new approach to the global economy is developing. By wedding the economy with the ecosphere, the world has the opportunity to develop in harmony with nature and as such be able to provide the resource base for this and future generations.

CHAPTER 2 AN ALTERNATIVE SOCIAL COMPACT

SUMMIT DECLARATION

The declaration of a world summit on social development must include at least five things:

- Recall and endorsement of other relevant resolutions, declarations, and principles;
- A consensus statement on the global dimensions of the problem being addressed;
- A vision of alternative social arrangements;
- An acknowledgment of the fundamental transitions required;
- A commitment to empower people to make the transition.

GLOBAL DIMENSIONS OF THE PROBLEM

- There are currently about 1.3 billion people living in absolute poverty and this figure is steadily increasing (Human Development Report, 1994). Every year 13 to 18 million people die from hunger and hunger-related diseases, most of these being small children (WCED, 1987).
- Income inequality is accelerating. Accordng to the UNDP, in 1970, the wealthiest 20% of the world's population enjoyed 73.9% of global wealth. Today that figure has risen to 82.7% At the same time the share of wealth received by the poorest quintile has fallen from 2.3% in 1970 to a dismal 1.4% today. The richest billion people command 60 times the resources of the poorest billion (Human Development Report, 1994).
- Mass unemployment in non-industrialised and industrialised economies is a serious and growing problem, the scale of the latter being a recent phenomenon. According to the ILO, 120 million people are officially registered as unemployed throughout the world. Unofficially, the figure could be as high as 820 million (30% of the world's labour force). In addition, another 700 million workers are considered under-employed. This means that even though these workers participate in some paid employment, they do not secure enough resources to meet a minimum standard of living.
- Two important causes of the current jobs crisis are global restructuring and technological change. Traditional hopes for full employment have rested on technological improvements, now however, technological developments have emphasized

increasing productivity and thus diminishing world labour requirements. At the same time, global competition and restructuring have led businesses to streamline their labour forces in order to turn the greatest profit. The net effect has been wholesale job losses. If things continue at the current pace, by the end of the first quarter of the next century, the vast majority of the expected population of 8 billion people will be neither producers nor consumers in the formal economy (Barnet, 1994). In the industrialized world, where consumption expenditures amount to nearly 65% of total GDP, this shift will have serious implications on the functionings of the economy (Clairmont, 1994).

- Unemployment is experienced disproportionately by youth. In some regions as many as 1/3 or more young adults are unable to obtain gainful employment. This leads to increasing alienation and disaffection for mainstream society and its values. Increasing violence and other criminal activities often follow suit (Barnet, 1994).
- Growing deterioration in the life support systems of the planet due to local and global environmental pollution and ecological degradation has led to the depletion of arable lands in a time when growing populations have put increasing pressures on the world food supply.
- The blind pursuit of economic self-interest has led to spiritual and ethical decline and the unraveling of the moral fabric of society.
- Conflicts are becoming an increasingly internal phenomena. Ethnic, social and political strife within nations is leading to civil wars around the globe. In fact, 79 out of a total of 82 such conflicts which took place over the last three years have taken place within the boundaries of a nation state (Human Development Report, 1994).
- Through-out the developing world, increasing debt burdens and balance of payment deficits offer serious constraints to economic growth and development. For example, in Sub Saharan Africa, the total debt load has quadrupled over the last 10 years and now equals 160 billion dollars. Much of the newly acquired debt is used to finance payments on old outstanding debts and therefore cannot be used to invest in much-needed human and physical capital.

- There has been a net flow of resources from the South to the North. Last year, developing countries gave the World Bank 2.76 billion dollars more than they received (Globe and Mail, 1994). This leaves few resources available for much needed social infrastructure like health and education. Often it is the poorest people of the developing world who are most seriously affected by resultant government cutbacks.
- There is an increasing disenchantment with the international system including the international trading system and the UN system. This seriously impairs the legitimacy of the process and negatively effects the ability of the system to reflect the ideas and values of the people it is trying to represent.
- Popular lack of faith in the political leadership of most countries is slowly eroding the democratic process. The ensuing apathy produces a serious barrier to the pursuit of change and the facilitation of the kind of paradigm shift required for sustainable development.
- There is a disturbing lack of a global sense of collective human purpose and identity.

VISION

For any vision of alternative social arrangements in human society to be long lasting, it must be based on the reality of our world as it is. This must begin with a recognition that human economy and society are embedded in and dependent on the natural system. The fundamental code of ethical human social behaviour must therefore be a pattern of relationships and activities among humans, and between humans and nature, which fosters the harmonious co-evolution of nature and society. Such a vision would be easily within our reach if the international community were merely to implement the agreements they reached in the recent UN Conferences on Environment and Development (Rio, 1992) and Human Rights (Geneva, 1993).

Secondly, it is important to work from within the framework of the "market system" and not against it. It is unrealistic to expect people as individuals and collectives to NOT try and maximise their economic well being. Any system which fails to incorporate this is doomed to collapse. At the same time, it is important to acknowledge that the marketplace, as it stands today, is supported by a menagerie of artificial props and supports which distort economic reality and as such does not take account for the full costs of production on both society and the environment. "*The only problem with capitalism is that it has not been tried yet.*" (Hawken, 1993).

It is also crucial to keep in mind that the crux of society is humanity itself, not the economy. Ultimately the economy should serve to facilitate human development, not ensnare and enslave human beings thus preventing them from reaching the fullest of their potential. Our vision is thus one where human beings are able, through the achievement of sustainable livelihoods, to lead enriching and fulfilled lives in synchroneity with the natural environment.

The upcoming UN Conferences on Population (Cairo, 1994) and Women (Beijing, 1995) will serve to sharpen that vision and define further pathways of achieving it. The Social Summit, above all, must embrace and reinforce such a world-view. A social order which is sustainable is implicitly consistent with a world which is sustainable. Such an order is characterized by respect and love for human spiritual and ethical values, wisdom, creativity, leisure, equity and social and economic justice.

FUNDAMENTAL TRANSITIONS

The global concept of "development" since World War II has been and continues to be based on the social and economic model of the developed or industrialized countries. Political leaders, development planners, academics and even NGO leaders in the South intuitively aspire to a state of development as characterized by the North. While some might articulate the folly of such a concept, in the next breath, they will speak of the "gap between the rich and the poor; the North and South or the underdeveloped and developed". This mimicry in development is so deep-rooted even in the multilateral "development" agencies that it is considered intellectually and politically naive to even discuss its inherent impossibility and imperialistic overtones, much less feasible alternatives. The first step we make towards vision must be the transition from the imposed and elusive goal of "development" and aspire to a world which encourages the flowering of individual cultures and human well-being as defined by these individual cultures. It is the sustainability of a wide diversity of cultures, human well-being within these, and the respect for such goals that must be the aspiration of the Summit.

Such a transition implies other fundamental changes. The pre-occupation of social science research, while bridging the "economic and social gap" within and between nations, must be refocussed and by an integration with anthropological and other approaches seek ways in which communities are allowed to define and meet their own aspirations within a country and between countries. The processes by which local communities and individuals within these communities are able to make a sustainable livelihood must become the focus of governments, NGOs and multi-lateral agencies. Such a vision is required in both the North and the South. Governments are known to be notoriously slow and inefficient at such transitions, but they must commit to it in the Summit. NGOs might have to take the lead and be supported by the research institutions and multi-lateral agencies.

The current view of society in which there are cores and margins has not been productive. It implies that the goal is to bring the "marginalised" to the social and economic levels of the so called "mainstream". It reinforces a view that the poor are objects to be helped. The recognition of the creative spirit, the adaptive strategies and possibilities of the so-called "margins" must be recognised for their intrinsic values and worth. These "margins" must be empowered to develop along their own lines in communal harmony with other social groups. The goal of social progress must be social harmony and human well-being, not defined by United Nations averages or by the life style of a particular group or group of countries, but by local communities and individual cultures; which develop and grow by exchange with other groups and cultures.

IMMEDIATE CHANGE

The transitions of which we speak in the foregoing section are part of a vision of a new social order. Therefore the question "But how do we achieve that?" will of necessity be difficult and even unanswerable at this stage. No apologies are required, for that is the nature of alternative visions. However, to the question "Is it realistic?", we dare to answer "Yes," largely because of the now undeniable energy and creativity of local communities. For too long these communities have been morally and materially disempowered.

At the same time, we recognise the need to address the more immediate or practical changes that are either current or necessary and feasible, and we do so briefly below and in succeeding Chapters.

REDEFINING THE WORLD OF WORK

As we approach the World Summit for Social Development (WSSD), it is apparent that many traditional and cultural conventions of work, e.g., the husband as the key "bread winner", the rural family as a unit of production, lifestyle and life cycle expectations, as well as other gender-based division of labor and patterns of residence are changing.

In both Northern and Southern countries, fundamental changes are taking place and are impacting upon the way in which jobs are created and people find gainful employment. Profound erosion is taking place in terms of expectations for life long careers, job security, and an ever improving quality of life. There is a global malaise in which expectations for gainful employment are impaired by spiraling debt, economic recession, and environmental degradation. People the world over are searching for job security.

Recently the G-7 convened a meeting on unemployment in Detroit to discuss unemployment and the global job crisis. One of the stark realities which emerged from the meeting was acknowledgement that the world of work has changed, and with it the social standards and expectations of an entire generation. Perhaps most alarming is the recognition that the job crisis translates into worldwide unemployment and is leading directly to social unrest, displaced populations, and a generation of new poor.

A parallel development is the emergence, since UNCED in 1992, of a nascent global consciousness which charges that our industrial work and consumption patterns are destroying the carrying capacity of the planet. As we approach the World Summit on Social Development, it will be necessary to give full voice to these two critical and interrelated issues. The world of work has been traditionally dependent upon an unlimited stream of cheap natural resources. As the living and non-living natural resources disappear, and their ecological niches are replaced with toxic waste, we see a two fold crisis emerging: the loss of sustainable employment and the ever expanding contamination of our environment. These two crises are universal in scope and are already affecting the physical and psychological well-being of people everywhere.

THE SCOURGE OF MASS UNEMPLOYMENT

According to the editors of the *Third World Resurgence* in the April 1994 issue entitled Workless – Mass Unemployment in the New World Order, "the world today is threatened by the scourge of mass unemployment. Not since the 1930's has this problem assumed such fearful proportions. The staggering scale of the problem can be gauged from a single statistic: about 30% of the world's labour force is unemployed".

In the United States alone, 10% of the population is estimated to depend upon food banks. Worldwide unemployment predictions present a foreboding picture for the future livelihoods of millions of people in both the industrialized and developing world. As Jeremy Rifkin indicates in his forthcoming book *The End of Work*, an estimated 90 million jobs out of a total of 124 million in the USA alone, are likely to be phased down or eliminated. Clearly, it is time to take a second look at the world of work, and to examine some of the underlying assumptions and causality which are affecting this massive global rupture in employment patterns.

Unfortunately, in today's world, there are no simple answers to shifting patterns of work. In July, 1994, the Government of Canada confirmed that fewer Canadians now work a normal work week of 35-40 hours. There has been an increase in those working either shorter or longer hours. This in effect has resulted in an increase in the inequality of earnings. In Canada, wage and income disparities exist among young and old workers in all industries and the gap is widening. The social impact of this phenomena is only beginning to be appreciated. Slowly workers are realizing that lifelong employment, job security, tenure, and pensions are social artifacts of the past – and that the workstyles of the future will be characterized by episodic, project based, and "quilted careers".

One of the realities of the recession of the 1990's is that the old "work order" approach or paradigm has changed. Public and private organizational structures are changing and strategies such as "work force adjustment programs" have become the order of the day as business and government struggle "to do more with less." Traditional values have also changed as have expectations for lifelong careers and job security. In Canada today, 25% of the total number of businesses are now home based. Displaced workers are forced to redefine their expectations and to undertake more direct control over the management of their livelihoods.

The job turmoil found in Canada is indicative of the escalating social "dis-ease" which is taking the industrialized and developing world by storm. No amount of structural adjustment, reduction of external debt and debt service can address the root cause of this malaise. The paradigm is shifting – and we must be prepared to address the consequences of our generation's lifestyles and consumption patterns which have occurred at the expense of the global environment.

Indeed, the present global crisis in employment is symptomatic of a larger environmental problem. Despite the vast wealth created by the world's $21 trillion economy, the quality of life on the planet is worsening for the majority of people. Ecological and social systems are in jeopardy and the health of our Earth is in decline. Moreover, essential life support systems are changing so rapidly, that in our lifetime alone, we have reached a major crisis in terms of being able to sustain the health of our planet.

The problem is not a lack of work to be done. Rather, there is a vast amount of work waiting for human attention, including building decent places to live, raising children, teaching one another, visiting, comforting, healing, making music, telling stories, creating art, inventing things and governing ourselves (Idris,1994).

As Barnet (1994) notes:

> "Until we rethink work and decide what human beings are meant to do in the age of robots and what basic economic claims on society human beings have by virtue of being here, there will never be enough jobs."

Remarkably, the linkage between environment and social progress has been an under-discussed topic in the preparatory arrangements for the World Summit for Social Development. Apart from modest references to Agenda 21 and the Earth Summit of 1992, the reality that social and biological forces are interdependent and inseparable, has not been fully recognized.

CHAPTER 3 EMPLOYMENT FOR SUSTAINABLE DEVELOPMENT

INTRODUCTION

In this chapter we discuss employment issues which are pertinent to poverty eradication and social integration. In particular we discuss the concept of sustainable employment from the perspective of enhancing the opportunities for each human being to make a living in a decent and sustainable manner. The interdependence of employment and the environment is also emphasized.

In illustrating this approach, we have drawn examples and principles from practical experience, mostly from the recent IISD Conference on Employment and Sustainable Development, which was convened in Winnipeg, Canada. While many are from Canada, we have selected those which have international relevance. The Case Studies themselves and related success stories are presented in Chapter 4. These examples illustrate key points based on substantive information, actual case studies, and feedback from conference participants.

Responsible environmental management practises are a prerequisite for sustainable economic development. Conversely, unchecked or unregulated economic development leads to further decline of environmental quality. Thus it is essential that economic development and ecological restoration are integrated to create a climate in which citizens can realize fair and equitable employment without compromising the quality of the environment.

The employment outlook for many Canadians and people the world over, has changed dramatically in recent years. The number of jobs available in agriculture, fisheries, forestry, mining and manufacturing has declined steadily. Concurrently, we have witnessed the adoption of new biotechnologies, changing market conditions, a degraded environment and depleted resource base.

The formal labour market is becoming increasingly polarized into so called "good jobs" and "bad jobs". Good jobs are those which are full-time, permanent positions which offer good wages and benefits packages, job ladders, skilled and meaningful work. On the other hand, "bad jobs" are typically part time or seasonal in nature, the wages are generally low and benefits nil. Such jobs are typically low skill, dead-end jobs. Many service sector jobs fall into the latter category.

A rapidly growing service sector has absorbed some of the displaced workers — often at a lower wage — but high levels of unemployment persist. The official unemployment rate in Canada hovers between 10 and 11%, but in some regions, e.g. in Atlantic Canada, persistent pockets of unemployment can be found in the 20 to 25% range. Underemployment is increasing and young people find it especially difficult to secure employment, even after years of post secondary education and training.

Fundamental shifts in the economy and accelerated technological change have left millions of people with a sense of powerlessness. Exclusion from the "world of work" breeds loss of self-esteem, discontent, and fear, often leading to behaviour that is self-destructive and which erodes the fabric of the family and society at large.

No single action will restore vigour to the economy and reduce chronically high levels of unemployment. The key is to focus on interrelated options that simultaneously promote economic, environmental and social gain. Sustainable development can provide prosperity over the long run. But achieving sustainable development, defined by the Brundtland Commission as "development which meets the needs of the present without compromising the ability of future generations to meet their own needs", requires a more integrated and long term approach than previous development efforts. Indeed, with the 50th anniversary of the World Bank and the International Monetary Fund (IMF), nations have an opportunity to learn from the "lessons" of the past and to recommend structural change and reform where needed.

SUSTAINABLE EMPLOYMENT CRITERIA

How are we to decide whether a job or employment opportunity is sustainable? This is the key question for individuals and communities who are making their own decisions. It is also critical for governments, which are trying to set the stage for long term sustainability. The criteria are not new: Each has been used before, however, sustainable development provides a basis for bringing them together in one analytical framework, and thereby finding acceptable pathways among apparently conflicting goals. The conflicts are more apparent than real, however, because in the long run, jobs must be environmentally and socially, as well as economically sustainable.

The impact criteria listed below are chosen in part because they are capable of being measured or estimated at the selection stage for a job category, or at the planning stages of a policy or project proposal.

ECONOMIC IMPACT

No job or work opportunity is sustainable if it is not a positive economic contributor. For individuals, companies or communities, a simple market test gives an immediate answer to this question: Is someone willing to pay enough for the good or service being produced to make it worthwhile? But this is an answer for today only. For the long term, people must look both to their own strengths and skills, and to the evolving needs of the market. For government policy, in particular, an investment orientation to employment policy is essential. What is the payback for an expenditure on training?

Specifically, does the employment in question:

- Provide needed products/services?
- Produce an acceptable return on investment?
- Provide a decent wage?
- Generate exports or substitute imports?
- Form part of a competitive or cooperative enterprise?

ENVIRONMENT IMPACT

Because ongoing economic activity depends on a functioning environment, we must consider how any activity affects it. Such assessments now are usually done at the project or site basis, and this will work when the particular work an individual is doing is part of a larger employment situation. But as employment and companies become more specialized and niche oriented, and as more individuals become (in effect) free-lancers, this site orientation becomes less relevant. Government and education policy makers must take this into account as they try to give individuals skills that will be useful in the future.

Specifically, does the employment improve the environment by:

- Helping to achieve environmental targets?
- Correcting or mitigating a problem?
- Increasing resource and energy efficiency?
- Decreasing the use and release of toxins?

SOCIAL IMPACT

For the individual, the social impact encompasses his or her relationships with the family, the community, and the individual's investment in their own future employability. Specifically, does the employment enhance the quality of life by:

- Promoting community survival, stability and development?
- Promoting family stability?
- Encouraging education and the learning of skills in demand, both for the employee and the family?
- Providing training that will lead to ongoing employment opportunities?
- Enhancing social equity?

Any individual employment situation can be evaluated in terms of questions like these. If the criteria are correctly chosen, they will allow both existing jobs and new opportunities to be measured for their sustainability. Measuring progress in sustainable development is a critical and difficult area. It requires bringing together data on social, economic and environmental progress so that decisions can be made taking all factors into account.

The above criteria demonstrate that progress can be made towards creating greater opportunities for employment, particularly when the three segments of society: the community, the government and the private sector join forces. The synergistic effect goes far in terms of generating energy, fostering political will, and sharing resources and talents.

The examples demonstrate that, while the mandate of the private sector is primarily to create wealth, the ethic to serve the public and promote social progress and welfare, should be a common value shared by government, the community and business alike. When these groups join forces, an enabling climate is created in which enterprise flourishes within the bounds of social and environmental responsibility. The cooperative or "tripartite" approach provides synergistic social and economic benefits to the community which go beyond the capabilities of any one group.

This tripartite approach should be further elaborated in the vision and concept of service to others which is articulated in the WSSD Copenhagen Declaration and Programme of Action. This approach promotes a dynamic enabling role for government along with the mobilization of community groups and active participation of private sector leadership.

Some key issues of employment and sustainable development to be discussed are summarized into the following issue areas:

- The culture of government;
- Modifying societal expectations;
- The changing reality of the ecosystem;
- The emergence of community governance and new partnerships among community, government and the private sector.

THE CULTURE OF GOVERNMENT

Alleviating poverty, putting people back to work, and spurring economic growth are essential to the future well-being of any nation. Governments must be committed to the provision of new skills and entrepreneurial training and must facilitate access to credit and viable employment opportunities. These are crucial components of any new government strategy.

Given current economic, environmental and social conditions and constraints, it is clear that new governmental growth strategies must focus on activities which will promote social development.

GROWTH STRATEGIES

- Minimize environmental impact by using energy, materials and resources efficiently;
- Restore degraded ecosystems and enhance biodiversity;
- Include marginalized members of society such as the poor, the unemployed, youth, women, minorities, First Nations and individuals with disabilities;
- Build on existing strengths and rectify weaknesses;
- Select activities which result in the greatest synergies and spin off effects;
- Provide needed goods and services in international and domestic markets;
- Build the local regional and national capacity required to adapt and respond to rapid change.

Although central governments are responsible for ensuring that the policy framework is in place, they cannot possibly control and deliver all the technical expertise, training, networking, and financing required to achieve a sustainable future. We recognize the need for more government leadership in these areas, along with more streamlined and coordinated practises. Jurisdictional overlap among federal or central government ministries, and between federal, provincial and municipal governments are major obstacles to kickstarting the economy and enhancing economic growth.

The "culture of government" needs to become more transparent. There is a tendency to have too many policies, programs, regulations, levels of government; all of which present barriers to achieving sustainable development. Intelligent delivery of goods and services can often be undertaken by the community in lieu of additional government infrastructure and without dependency on government.

In many countries, people have abdicated their responsibility to manage their affairs and livelihoods. There is an expectation that government will provide leadership in managing the change process. In working towards sustainable development, communities must assume greater responsibility for managing their local economies and infrastructure. The adage of government carrying for citizens from th "womb to the tomb" is no longer relevant.

There is nevertheless an important role for government to serve in education and training; namely to facilitate capacity building of community resources, technology transfer, information technology, electronic tools, geographic information systems (GIS), strategic and fiscal planning tools, market niche development, value added advice, and export enhancement.

One of the major conclusions of the Winnipeg conference was the necessity for individuals and communities to "take back" best practice and to acquire or re-acquire control of their livelihoods. Sustainable development cannot be achieved with conventional top-down bureaucracy. It is a circular process in which the community, government and industry join forces to promote sustainable economi growth and social progress. This new tripartite partnership is a logical extension of the community roundtable consultation process promoted by the Brundtland Commission in the late 1980s. While this is a predominantly western model for "process based" development and resource co-management, the principles hold promise for other nations. In return, the processes used in these nations can feed back into the round table approach.

By fostering the development of innovative partnerships, practices, technologies, networks and social programs, governments can help revitalize their economies. A sustainable development policy framework that provides leadership by defining strategic goals and objectives and putting the right signals and mechanisms in place is essential. In Newfoundland, Canada, for example, the provincial government Economic Recovery Commission has engaged in a joint effort with the public and private sectors to identify 14 new growth sectors. By assessing local strengths and weaknesses, undertaking resource surveys, and fostering community dialogue, these areas have been selected as priority targets for development. The Newfoundland example illustrates the importance of identifying profitable opportunities, integrating approaches to realize maximum potential, and forging new partnerships to mitigate potential difficulties. In this example, special care has been taken to rely on a range of stakeholders representing broad knowledge bases, to strive for synergy, and to take care not to foreclose options.

CHANGING SOCIETAL EXPECTATIONS

Coming to grips with the changing nature of work is essential to fostering sustainable livelihoods. Equipping workers with the knowledge and skills necessary to adapt, learn and become more flexible and self-reliant is the paramount objective. Equally important is the recognition of income generating opportunities and a support structure that empowers and includes individuals with varying levels of skills and knowledge. Changing employment trends pose additional challenges to training and social assistance programs. Non-standard work, including part time and multiple jobs, and partial year, temporary, seasonal, and self-employment have risen dramatically in the past decade. In Canada, almost 60% of the jobs created in 1993 were part time, and the majority (70%) of these were filled by women.

As discussed previously, our societies are facing major shifts in focus and attitude. Conventional institutions and units of production such as the family are experiencing severe stress due to rural urban migration, disease – e.g., HIV/AIDs, war, and changing labour patterns. The search for security is a daily struggle for millions of the world's urban and rural poor. In developing countries, unemployment and workforce downsizing are creating stress, social instability and new classes of poor. Middle class expectations of lifelong and universal jobs, gainful employment, retirement pensions, are also changing. The world of work, indeed the essence of work, has changed. Societal expectations must be reoriented to help individuals manage the change process.

Economic development that helps regenerate environmental quality and ecosystem health, will value human well-being and ecological integrity above the material throughput and industrialization strategies of the past. Adopting such an approach requires a new way of thinking about objectives and expectations. The linear input-output models of the industrial world are not compatible with biological systems. First Nations and many Eastern cultures have always recognized the need to view life processes in a circular fashion. Western industrial systems however are premised upon a linear view of the world; high input matter and energy consumption; high throughput and consequently massive waste generation.

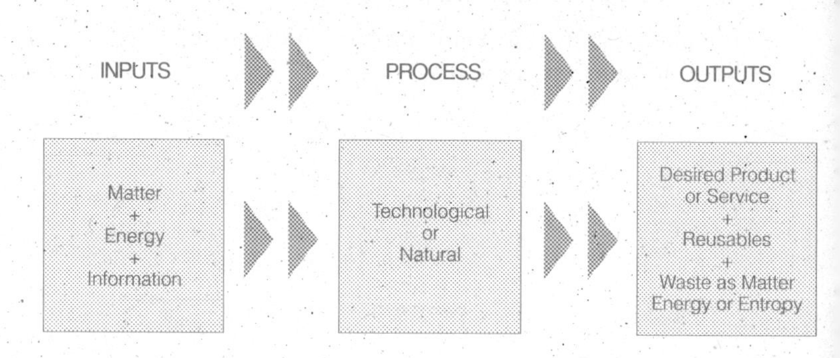

Linear Design
Source: Singh, Naresh. (1994). *Ecodecision*, Vol. 11, p. 58-59.

Assumptions of a world operating as a linear system are a direct violation of the first Law of Thermodynamics; from which it becomes obvious that whatever is taken from the environment will be returned, albeit in a different form. Typically, we take materials and available energy and return waste and entropy. When we forget that natural ecosystems must metabolise or cycle these outputs so that they can be useful to us again as inputs then we ignore the obvious cyclical design of our world and live a make-believe unsustainable lifestyle. Such a fundamental misconception alienates human society from nature and sows the seeds of increasing social and natural chaos.

> "The only way to increase employment is to decrease throughput, how to move from a least price to a least cost is the industrial question of this and the coming century. It costs less to take care of the Earth in real time and it takes more people to do it. The technologies required to reduce throughput and restore natural systems have largely been invented. Now we need to put people back to work using them." (Hawken, 1993).

Providing long-term sustainable livelihoods to the vast range of the world's population is a growing challenge. As governments around the globe grapple with unemployment, massive and rising deficits, and continued environmental decline, a small but growing number of enlightened policy makers are realizing that sustainable development involves a cyclical, cooperative and consultative process. It is commonly agreed that sustainable activities:

- Meet essential human and societal needs;
- Protect or enhance environmental systems;
- Generate a profit or reduce costs;
- Integrate resources in a circular fashion compared to the linear production approach.

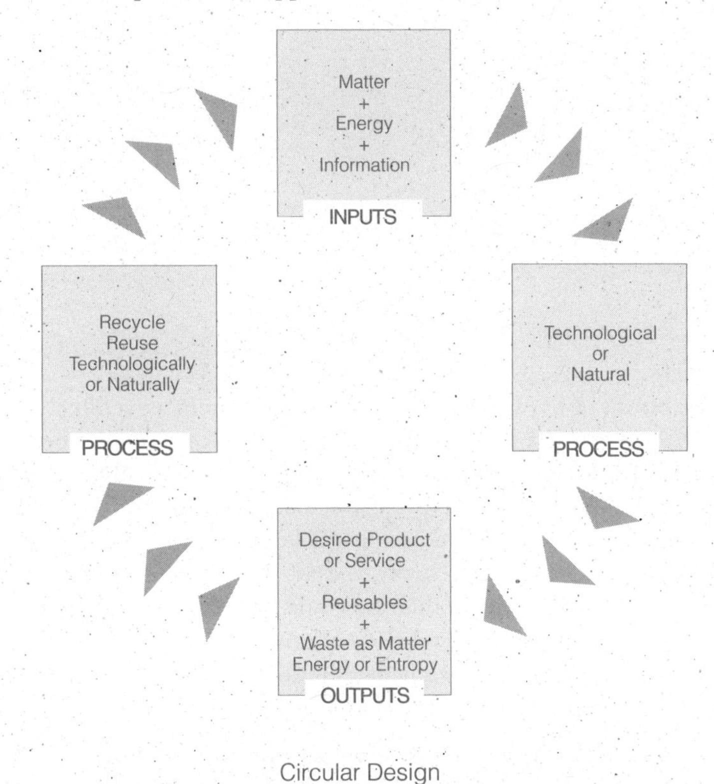

Circular Design

The pursuit of sustainable activities by definition, does not foreclose options for future generations. In fact it provides a setting for the creation of more choices. By respecting inter-generational considerations, sustainable development concepts find favour with all nations and cultures. The respect for traditional knowledge and wisdom is a principal tenet of this approach.

Accelerated technological change, increasingly fluid international markets, and dramatic economic restructuring are influencing the ability of companies, communities, and individuals to manage the marketplace. Changing environmental and social values are also exerting new pressures on economic choices. All of these exert pressure on societal expectations and aspirations for a better quality of life.

There is a need to shift the emphasis on resource exploitation from production driven to market driven. In certain cases it may be necessary to create new market demand, e.g. to create a market for a specific product. As the quality of product affects the potential for "value added", it becomes imperative that production integrity and excellence set the standard.

New opportunities are emerging for communities to undertake self directed management in areas of value added product modification, niche marketing, recycling, waste product utilization to name a few. Communities are beginning to undertake to build upon their resource strengths and expertise, and to develop strong "eco-efficient" domestic and export markets.

Some of the sustainable activities identified as significant potential employment generators in Canada and which might be relevant to several other countries include:

- Comprehensive building retrofits to upgrade energy, water, waste, and indoor air quality systems;
- Increased reliance on aquaculture to satisfy a growing demand for fish and seafood products, particularly export demand;
- Provision of quality child and elder care services;
- Improved management and harvesting of forestry, agriculture, and fishery resources, including expanded tree planting in urban areas;
- Adding value to bulk commodities of all types – agricultural, forest products, fish, minerals and chemicals;
- Collecting, sorting, and processing recyclable materials;
- Development of technologies that reduce the consumption of energy and materials; diminish the effluent and pollution generated from economic activities; reduction of waste streams; restoration of degraded environments;
- Tourism development based on sites of environmental, cultural, and historical significance;
- Extending the "information highway";
- Computer literacy training;

- Distance learning services;
- Geographical information services (GIS) resource mapping;
- Natural resource inventory – e.g., coastal and ocean floor mapping;
- Development of alternative transportation systems – bicycles, mass transit, solar vehicles;
- Development of renewable energy and cogeneration technologies – biogas, solar, wind, wave energy;
- Development and marketing of technologies and practices required to comply with international environmental agreements re: ozone depletion, climate protection, biodiversity protection.

Taking the above points into consideration, the goal and expectation of sustainable development is the manufacture, and delivery of goods and services which satisfy human needs, which are competitive, and which result in a dramatic reduction in resource utilization and waste generation. Many of the innovative technologies, products, services and designs needed in the new economy have been developed and are ready for introduction or wider adoption in the market place. Governments should put into place the appropriate policies, institutions, financing mechanisms, education and training infrastructure which will accelerate their use.

Today, small business are becoming increasingly important "engines" of the community-based economy – particularly as sources of innovation and employment. In Canada, almost 25% of the total number of businesses are now home based. The era of the "electronic cottage" is revolutionizing the way Canadians work and interact.

From experience in many countries, we have learned that smaller scale initiatives are less risky, tie up less capital, provide greater flexibility in changing circumstances, and because of the multiplier and spin-off effects they have the potential to create more economic activity and employment. For example, energy conservation and small-scale renewable technologies for generating electricity may provide four times the number of direct jobs per dollar invested than do large generating projects.

WORKING WITH THE ECOSYSTEM

Economic activities that impair the functioning of natural systems or that exclude people from decision making and resource co-management, are not sustainable. The collapse of the commercial fishery in Atlantic Canada, persistent pollution in the Great Lakes basin and the appalling conditions of First Nation communities, are all symptomatic of development strategies gone wrong. Recognizing the causes of current problems and resolving to redress them is the first major step toward creating positive change.

While predicting the future becomes more difficult in times of unsettled rapid change, there are many guide posts to help us. Canada for example, has committed itself to a rapid phase out of chlorofluorocarbons, to reducing CO_2 emissions, to preserving its biological diversity, to preventing or restricting pollution of all types, and to restoring the ecological integrity of its forests and watersheds. Through federal and provincial Roundtables on the Economy and the Environment, extensive public awareness has been generated.

It is now recognized that sustainable economic activities can create profits by using resources efficiently, by maintaining the integrity of natural systems, by fully tapping and rewarding the knowledge, skills and creativity of individuals and communities, and by providing products and services demanded in the market place. Fostering these activities requires an educated and well-trained labour force, efficient government policies, a heightened spirit of self-reliance and entrepreneurship, and an investment climate which embraces the concepts of social and environmental responsibility.

Progressive economic development values traditional knowledge, the quality of life of community residents, and principles of restorative ecological management. This approach or philosophy of development also places priority on intergenerational needs. The bottom line for business should no longer be profit alone, but social and environmental responsibility — and an assured future for the next generation. This is an alternative social contract model for the Twenty First Century.

Quite simply, the imperative for sustainable development is one of the most important messages which can be imparted to the delegates of the World Summit for Social Development. It is only very recently that we have begun to understand the critical relationship and interdependence between industry, the environment, and social-well being. It is essential therefore that the guiding principles of "restorative ecology and economics" be carried forward, championed by the delegates, and integrated into the Copenhagen Declaration and Plan of Action.

COMMUNITY GOVERNANCE

The role of community economic development is to increase productive linkages within the local community and decrease "leakages" or loss of energy and resources. The purpose of these strategic efforts is to achieve synergies, include and empower the greatest number of people, and strengthen economic competitiveness. Organizing alliances to mobilize people, finances, technical expertise and real property are the tools.

Communication technologies are enabling individuals and small firms to access information, develop fax, modem, electronic mail networks and alliances, and market products and services in entirely new ways. Self employment assistance programs and business development centres for small business startup, i.e., incubation centres, are on the rise in Canada.

With respect to greater community governance, fundamental government policy shifts are required to encourage the development of a restorative economy. Fiscal policies that reward stewardship, collective income generation and savings are preferred in lieu of policies that encourage consumption and spending. Loan guarantees and grants that assist small, appropriate local initiatives generate more synergies and spin-off effects. These small community based businesses strengthen regional economies more than financial assistance to megaprojects which often utilize external consultants and firms to deliver goods and services.

It is widely accepted that a major increase in human population and consumer activity will occur during the next 50 years. Given that the natural capital, (i.e., the living and non-living resource heritage of the planet) is in jeopardy, new ways must be found to manage the change process and to assist the world's poor to achieve self-reliance and security. It is no secret that conventional development assistance policies and programs of multilateral and bilateral aid programs have not been very effective and in many cases have resulted in exacerbating poverty and environmental degradation. Perhaps most devastating is the trend of creating greater dependence on external systems which deliver goods and services at the expense of local self-reliance and the local resource base.

Many of today's most common technologies, such as the personal computer, video cassette recorder, microwave oven, fax machine, and compact disc players did not exist 20 years ago. Earth observation satellites, CD ROM technologies, computer assisted design programs and a host of other advances are revolutionizing the way in which we conduct our work. Isolated communities should be given the opportunity to evaluate the potential of these communication tools. It is believed for example that vast majority of the research findings and data on Africa are found in institutions outside of the African continent. Enabling African research centres, universities, libraries and businesses to assess the potential applications of the "information highway" should therefore be a high priority. Communication linkages are the essential prerequisite for establishing sustainable development programs.

The continuing destruction of natural capital must be halted. The best hope and practice is to encourage individuals and their communities to take charge of the management and governance of their livelihoods. Targeting action for people who are disadvantaged socially and economically is not easy. The elimination of physical insecurity and the development of "sustainable" societies are two of the greatest challenges today. The development of a framework in which communities and governments assume responsibility for co-managing the environment and global commons (land, air, and sea) is a great step forward. This will require building new partnership and communication linkages on an unprecedented global scale.

Through the WSSD, opportunities exist to advance approaches which challenge the status quo and which offer alternative methods for integrating the poor into the social and economic development process. The framework for sustainable development which advocates that the leaders of private sector enterprise have a major role to play, in tandem with government and NGOs, in employment promotion, must be forwarded.

Community economic development approaches endeavour to strengthen all component parts of the community. The process is essentially one of devolution of power and financial resources from central government authorities to bonafide community groups. The focus of such programs empowers marginalized groups, and in particular addresses the needs and aspirations of the disadvantaged.

One of the key characteristics of this approach is the emphasis on "horizontal" as opposed to "vertical" decision making and communications. This methodology represents a departure from conventional models in which government plays the major role in terms of the delivery of goods and services. The process requires a substantial period of orientation, education and training.

It is apparent that many community groups and government agencies are presently experimenting with methodologies to implement the participatory model. These efforts to share power and resources represent a major breakthrough in our attempt at poverty alleviation and employment enhancement.

In Canada, much of the sustainable community economic development theory and practise has its antecedents in a social movement which began in Antigonish County, Nova Scotia, during the depression years of the 1930's. A grass roots adult education movement, known as the Antigonish Movement posited criteria for a development process which includes "putting people first", group action, needs assessment, structural and community transformation, and the utilization of ecological resources in a sustainable fashion. The Antigonish Movement represents the "best practice" in terms of community, industry and government partnerships.

EXAMPLE: THE ANTIGONISH MOVEMENT

In terms of "best practice", the Antigonish Movement principles were adopted by the Coady International Institute of Nova Scotia, and shared worldwide. Specifically, the Coady International Institute (CII) undertakes to promote:

- Growth in the personal capabilities of the disadvantaged;
- Collective empowerment through organization, transformation of societal structures which regulate lives and livelihoods;
- Transformation of cultural standards which constitute the normative blueprint for the way society is organized;
- Enhancement of the ecological resource base upon which communities depend for physical sustenance;
- Targeting for action the people in their societies who are relatively more disadvantaged socially and economically. These people should include unskilled and unorganized workers, entrepreneurs and marginalized peasants, unemployed youth, slum dwellers, indigenous and other minority groups. Women, by reason of their subordinate status within these groups and generally in society, should be given special status;
- Eliminating physical insecurity among the disadvantaged and facilitating growth of their human capabilities. Guaranteeing food security should be essential in this process;
- Transforming societal structures which deprive the disadvantaged of equitable access to public benefits and opportunities. The principles of social justice should constitute the normative framework for this development;
- Eliminating cultural practices which act as constraints on human and institutional development; nevertheless, the positive features of cultural traditions should be respected and protected;

- Enhancing the natural environment through ecologically sustainable development practices. Methods of agricultural production and production of energy should be priority concerns.

Source: Coady International Institute. (1990). "Policy and Program for the 1990's", pp.1-7.

CHAPTER 4 CASE STUDIES

INTRODUCTION

The purpose of this chapter is to provide examples and case studies which illustrate opportunities for achieving employment and sustainable development. The examples demonstrate the potential for addressing technological shifts, appropriate technologies, and enhanced North-South linkages. The majority of these examples were derived from the International Institute for Sustainable Development Conference on Employment and Sustainable Development, June 23-26, 1994, in Winnipeg, Canada.

The case studies and examples presented in this Chapter were extracted from the Canadian experience. It is important to note, however, that virtually all of the ecological and economic constraints listed in Chapter 3 apply to both industrialized and developing nations. The common denominator or characteristic in terms of poverty is the lack of control over one's environment and a lack of options.

The Canadian International Development Agency describes poverty as "the underdevelopment of human potential". The relationship between impoverishment, poverty and empowerment in the North and South have already been described (Chapter 3). The "underdevelopment" of individual and community potential remains a common challenge to both Southern and Northern countries. However, strategies for sustainable employment and social development will have to take the individual country profiles into consideration. Social development and employment strategies must address the context of poverty and social insecurity.

The concept of sustainable livelihoods provides new opportunities, scope and magnitude for understanding the linkage of employment to social progress. On the one hand, the concept identifies potentially unsustainable practises while on the other, new perspectives on the role of business, competition, aboriginal philosophy, and infrastructure can be examined.

The Canadian experience of National Roundtables on the Economy and the Environment (NTREE) provided a useful forum prior to UNCED for bringing together the competing interests to address common goals. This process is ongoing and has helped to create a climate of consultation in Canada in which agencies such as IISD play a major brokerage role. Growth areas including economic sectors such as forestry, fisheries, ecotourism, environmental industries, can be both profitable and sustainable. The conference also concluded that the challenge is not to create change, but to steer change in the right direction. At present, the growth areas in the Canadian job market are as follows:

- Transport equipment and systems;
- Forest management technology;
- Fisheries management/post harvest technology (value-added);
- Value-added processes for all natural commodities;
- Eco-tourism;
- Environmentally advanced technology – geographic information systems (GIS);
- Education, training, skill development.

In the following section we present a number of selected sectoral areas and case studies to illustrate sustainable employment concepts, barriers to success and achievements in the areas of forestry, land stewardship, environment industry, conservation industry, resource based products, indigenous communities, community renewal, aquaculture, building retrofits, value added industry, transportation and coastal resources.

FORESTRY, LANDSCAPES AND HABITAT

Some of the key principles of sustainable forestry are as follows:

- Management should be based on the development and maintenance of a healthy forest ecosystem, including non-commercially valuable species, rather than on the production of purely marketable, even aged stands. In this way, the natural cycle of forest decomposition and regeneration is preserved.
- In accordance with this, management units should be determined on the basis of ecologically determined boundaries, or, much greater cooperation and coordination must take place among public and private landowners in terms of their management practices. Sound management on one patch of an ecosystem can be undone by poor management in another patch.
- Forest management must be economically and socially viable, as well as ecologically sound. All three of these considerations are necessary conditions for forest sustainability, and the neglect of one will lead to the collapse of the entire system. Social viability involves the incorporation of local knowledge and skills, and the involvement of local populations in decision making.

- Maximizing local "value-added" is vital to increasing the economic incentives for sustainable forestry and to creating employment.
- The forest industry must be managed in a cyclical manner, and include the growth and stewardship of forests rather than simply harvesting and processing.
- New employment opportunities exist in improved natural resource accounting and mapping, in geographic information systems, energy cogeneration, and the diversification of products.
- Much greater understanding of forest ecosystem functioning and response is needed before sound, sustainable forest management practices can be established. Practices will inevitably differ according to the type of forest ecosystem, e.g. tropical versus temperate forests. Employment opportunities exist in furthering our knowledge of forests.

The Wildlife Habitat Canada organization recommends that "forested and agricultural landscapes be approached somewhat differently. In the first, the forestry sector needs to assume responsibility for a natural landscape approach to forest management. In the second, funds for employment creation could be used to create community-based centres charged with the identification and implementation of conservation projects of economic benefit to the community...". A "natural forest" based approach to forest management in Canada would produce the following benefits according to Booth, Boulter, Neave, Rotherham and Welsh (cited in Neave and Girt, 1994):

- Better protection of water systems, a secured supply of forest in various successional stages, including old growth, and a secured range of habitats for wildlife;
- No loss of net returns from logging activities as improved quality of time would offset any increased harvesting costs;
- Greater labour inputs per quantity of timber harvested but lower capital costs.

ECOSYSTEM APPROACH IN BRITISH COLUMBIA

With the release of *Progress Report #2: Review of Current Forest Practice Standards in Clayoquot Sound by the Scientific Panel for Sustainable Forest Practices, British Columbia,* in May 1994, a major step was taken by a Canadian provincial government to acknowledge the appropriateness of ecosystem-based management within Canada. The Scientific Panel was appointed by the provincial Cabinet in 1993 as part of its strategy to address worldwide concerns over current forest practices in British Columbia.

The panel report is historic. This is perhaps the first time (in Canada) that a team of highly credible scientists with a wide array of expertise has been called in to assist both forest managers and politicians to find a way by which forest harvest activity can be done sustainably. The findings of the Panel clearly demonstrate that significant, major changes in current forest practices must occur before we can consider forest harvesting activity to be sustainable.

"The general findings and recommendations (of the Panel) emphasize changes required in both the philosophy of forest planning and management, and the way that forest practice standards are created and applied. The Panel outlines the action required to make the transition from management of forests for products to management for sustainable ecosystems" – the point being that the transition in forest practices has not yet occurred.

The Panel found that "current standards do not recognize sufficiently the physical and ecological connections among terrestrial, freshwater, and marine ecosystems. Current standards do not effectively integrate ecosystem and cultural values. Nor do they address requirements for ecosystem sustainability, harmonious stewardship of all resources, and the needs of future generations." Current standards, "collectively, do not prevent the loss of biodiversity, degradation of terrestrial and aquatic environments, and damage to First Nation's heritage sites and areas. Nor do they ensure the restoration of ecosystems damaged by past development activity".

The Panel's report is an initial blueprint which could provide evidence of Canada's willingness to make serious progress toward application of an ecosystem-based management approach in its forests.

Source: Patterson, Doug and Robert Nixon. (1994). "Employment and Sustainable Development in Forestry: The Ecosystem-based Determinant Increased Complexity in Forecasting Employment Trends." Eco-Forestry Institute, Victoria, BC, Canada.

Ecosystem based forest management which accounts for a holistic relationship between ecosystems and cultural values, is slowly gaining favor with a number of Canadian experts.

ENVIRONMENT INDUSTRY IN CANADA

Background

The environment industry provides environmental management services, including waste management, resource conservation, pollution control, planning and the provision of information.

Significant employment opportunities exist in evaluating building performance, advising building owners on cost-effective upgrades, manufacturing, marketing and selling energy efficient products, and installing improved technologies.

EXAMPLES OF THE ENVIRONMENT INDUSTRY IN CANADA

In British Columbia, newly developed Building Environmental Performance Assessment Criteria (BEPAC) are being used to evaluate the environmental performance of new and existing office buildings. Building performance is assessed relative to ozone layer protection, environmental impacts of energy use, indoor environmental quality, resource conservation, and site and transportation. The assessments are a management, strategic planning, design, communication, and public education tool. Investments expected to result from the assessments will generate demand for environmental products and services and provide new jobs.

In Ontario the Ministry of Environment and Energy and Jobs Ontario recently launched two initiatives to promote the greening of homes and industries. With an investment of $41.8 million over the next three years, the program is intended to stimulate demand for environmental goods and services, trigger private sector spending and investment in green retrofits, and create 12,000 jobs. Both programs are expansions of existing pilot projects.

The Home Green Ups program, funded at $26.4 million, will promote household energy and water conservation and waste reduction in an estimated 250,000 homes in 23 Ontario communities. Pilot projects in seven Ontario communities have already resulted in 4,000 home audits.

The suggested and installed improvements have reduced water consumption by 25%, energy consumption by 15%, and waste generation by 15%. They generated about $5 million in municipal and private investment and created about 150 new jobs. The audits often include on-the-spot installations of energy and water efficient products. Auditors also provide referrals to approved local contractors and equipment suppliers.

The Green Industrial Analyses and Retrofits program, funded at $15.5 million, is a cost-sharing approach expected to assist an estimated 90 industries conduct 200 green retrofits. In step one of the process, the Ministry will provide up to 75% of the cost of a green industrial analysis prepared by a consulting engineer. The audited company is then encouraged to make the investments required to achieve the identified savings. For retrofit programs with a payback of longer than 1.5 years, the Ministry will cover up to 30% of the cost with a maximum contribution of $300,000.

At the federal level, a plan to upgrade and retrofit 50,000 government owned buildings could generate thousands of jobs and $1 billion worth of business for the construction industry.

Improved and expanded training is required for youth, the unemployed, and upper age trades people in order to perform the thousands of building audits and efficiency upgrades needed across the country. Participants at the Employment and Sustainable Development Meeting agreed that developing nationally recognized training standards and certification programs is essential. Integrated training across a variety of disciplines must be coupled with on-line access to information on the latest building technologies, codes, and practices.

Much of the renovation and construction work done in Canada is currently performed by individuals who get their training on the job. Lack of formal educational centres, electronic and distance learning options, and established apprenticeship programs hinder entry to the building professions and make adoption of the latest design and retrofit techniques haphazard at best.

Glenn McKnight of the Energy Conservation Society of Ontario recommends the establishment of self-directed learning centres, staffed by teams of knowledgeable trades people, to fill the gap. Based on a coaching, rather than instructor, approach, the centres could provide trainees with basic education in the building trades and guide them to on-line sources of the latest information in their area of interest.

In short, industry accomplishments are as follows:

- Environmental protection industry presently employs 60,000 to 70,000 people in Canada;
- Many pilot projects are underway to employ thousands of high school graduates in the environment industry through

practicum study partnerships between industry and community colleges;

- As a growth industry, new companies are emerging to address problems of erosion care, deforestation, pollution;
- New job creation is centred in agriculture with an environmental context e.g., environmental on-farm planning; export markets for environmental technology and consulting services; marine biotechnology, building retrofit programs e.g., waste, energy, deposit/refill systems; integrating environmental considerations into virtually every industry;
- New community initiatives include commercially viable recycling operations, research into energy efficient practises, national and international internships and "disassembling", e.g., developing new products from wastes such as carpets from old tires;
- The building "retrofit" industry is an environmentally sensitive growth area, especially in terms of domestic/residential dwellings and provides an alternative to new construction and natural resource utilization.

EXAMPLE — JOBS IN ENERGY CONSERVATION IN ONTARIO

According to the Energy Conservation Society of Ontario, Canadians have the dubious honour of being the greatest per capita consumers of energy in the world, with 25 - 33% of the energy consumption taking place in the heating and cooling of homes. This is 50% more than countries such as Sweden which have a similar climate and population.

The Energy Conservation Society of Ontario has identified key/new job opportunities emerging out of a concept called storefront Energy Saving Corporations, commonly called Energy Service Companies (ESCO's). A potential growth of 100,000 self-financing jobs can emerge in the Canadian economy, primarily in the energy efficiency, water conservation, and waste management sectors. This will provide a much needed value-added service where in a competitive world economy we are forced to reduce the cost of operating business by making buildings more energy efficient.

Source: McKnight, Glenn. (1994). "Green Enterprises: Energy Retrofitting." Energy Conservation Society of Ontario Oshawa, Canada.

Energy conservation represents a new and dynamic sector of the economy which is capable of creating thousands of new job opportunities.

RESOURCE BASED PRODUCTS FROM THE NORTH AND ABORIGINAL COMMUNITIES

Case Study Context

- In the Canadian North, the majority of the population is under 25 years of age. Although the population is relatively small, extensive immigration from the southern reaches of Canada has been triggered by mineral exploration. Capital investment tends to be focused in the southern parts of the country, and there is uncertainty as to how the resolution of $1.5 billion in aboriginal land claims will impact upon non-renewable resource extraction.
- Traditional First Nations knowledge and spirituality also impacts upon entrepreneurship. New growth industries are emerging in terms of eco and cultural tourism.
- Sustainable hunting and trapping industries are emerging.
- New industries based on northern resources, e.g., wild fruit processing, seasonal fish plants. Potential exists for an international aboriginal trading network utilizing electronic mail for securing worldwide markets and commodity exchange.
- Northern cultures are experiencing a spiritual renaissance and process of healing after decades of exploitation and isolation. Potential jobs exist in terms of exporting northern aboriginal knowledge and spirituality.

CASE STUDY — SUSTAINABLE RESOURCE MANAGEMENT IN AN ABORIGINAL COMMUNITY

The APIKAN Indigenous Network notes that in the past millennium, Aboriginal economic activity has shifted from dominance in the primary sector to dominance in the service sector – and Aboriginal peoples have been pushed to the margins of the economy, even in their own territories.

The Inuvialuit Final Agreement (IFA) established five renewable resource co-management bodies. The bodies include: Environmental Impact Screening Committee, Environmental Impact Review Board, Wildlife Management Advisory Committee for the Northwest Territories, Wildlife Management Advisory Committee for the North

Slope and Fisheries Joint Management Committee. According to Inglis (cited in Brascoupe, 1994), "while the IFA has not created a fully autonomous system, it has in a sense made agency arrangements with government: for the Inuvialuit to perform some of the tasks previously undertaken by government, and for the government to perform some of the Inuvialuit's functions by means of its agencies.

A result of the work of these bodies is the Inuvialuit Renewable Resource Conservation and Management Plan which sets out a long-term strategy for local community development of fish, wildlife, and other renewable resources. In addition, the Renewable Resources Development Corporation was created in 1990 to develop viable economic ventures including commercial fisheries, tourism, and restaurants featuring northern foods.

Similarly, the Shuswap Nation Tribal Council in western Canada has established an institute which will develop plans for habitat restoration on a regional scale. The plan will integrate forestry, mining and agriculture. The Tribal Council is using an electronic bulletin board to share information and the system will be made available to other indigenous communities along the Fraser River Valley.

Aboriginal Peoples may come full circle by becoming major players in resource restoration and sustainable resource harvesting. This would enable Aboriginal Peoples to regain control and management of the traditional territories and would be a major contributor to Aboriginal employment in the future.

Source: Brascoupe, Simon. (1994). "Sustainable Cultural Development: Sustainable Development in the Past and Future of Aboriginal Employment in Canada." Apikan Indigenous Network, Ottawa, Canada.

Through resource restoration and sustainable resource harvesting, aboriginal peoples can regain control and management of traditional territories.

COMMUNITY RENEWAL

The focus here is on taking stock in rural areas, the need to engage communities in the process of developing sustainable economic development plans; to attract business investment to the community; and to forge new partnerships with business and government.

General considerations:

- Communities need facilitation support to identify skills and resources, and to prepare inventories of human and natural resources;
- Communities are undertaking the development of action plans and inventories of resources; there is considerable interest in electronic linkages to outside centres of knowledge and expertise;
- Community needs and obstacles to progress need to be identified – "visioning" and needs assessment exercises are the first step to setting goals;
- Communities are interested in local control and decision making mechanisms such as community "trust funds", and bridge funding;
- Communities see rural food-producing jobs as a high priority and are looking for ways to strengthen the entrepreneurial expertise in the community.

CASE STUDY — COMMUNITY RENEWAL

Community economic development (CED) is a comprehensive, multifaceted strategy for the revitalization of community economies, with a special relevance to communities under economic and social stress. Through the development of organizations and institutions, resources and alliances are put in place that are democratically controlled by the community. They mobilize local resources (people, finances, technical expertise and real property) in partnership with resources from outside the community for the purpose of empowering community members to create and manage new and expanded businesses, specialized institutions and organizations.

RESO (Regroupement pur la Relance Economique et Social du Sud-ouest de Montreal) provides both employability services and other services to businesses. The organization is also involved in issues related to land use, development of infrastructure, and promotion of the area. It experiments with innovative approaches to reaching those hardest to reach such as chronically unemployed youth. Formetal, a successful training business in the metallurgical field has come out of this effort.

RESO has a mandate for the economic and social renewal of southwest Montreal, an area which has suffered continuous industrial decline over the last 20 years. In some neighbourhoods,

50% of the population is on social assistance and unemployment reaches 35%.

RESO is a membership based organization. Its board structure consists of four representatives elected by the community movement, two elected by trade unions, one from big business and one from small business. Over the last two years, RESO has trained over 1500 poor people. Training investments are continuously becoming more effectively linked to the local labour market. RESO has provided technical assistance to over 200 businesses in the last two years. In the height of a recessionary period, SW Montreal has, for the first time in 20 years, halted the decline in its manufacturing base.

Source: Lewis, Mike. (1994). "Community Economic Development Lessons from the Trenches: Directions for the Future." Centre for Community Enterprise, British Columbia, Canada.

Small community-based organizations are best able to assess the capabilities and needs of the community and as such directly facilitate local development efforts.

AQUACULTURE

Background

- The aquaculture industry covers a wide scale of operations from large multinationals, to small "mom and pop" operations. The culture of fish implies human intervention in the rearing process and individual or corporate ownership of the stock. In Canada, a wide variety of fin fish and shell fish species are currently farmed.
- The aquaculture industry is viable and operable all year long.
- Opportunities for aquaculture exist mainly in rural and coastal areas; the business structure can be quite diverse and is usually export oriented.
- This industry is characterized by very high growth rates. In Canada, the industry employed only 200 people in 1984. By the year 2000 it is expected that 12,000 new jobs will be created for a total value of $677 million of product.
- Aquaculture products are the fastest growing commercial fish products in Canada and are replacing traditional wild harvest fisheries, particularly for species such as salmon. The critical success factors in aquaculture are: government commitment;

regulatory reform, e.g., access to production sites, therapeutic and inspection services; strong export promotion, industry coordination; access to capital, training and education, and enhanced public awareness regarding the low cholesterol and health qualities of the product.

- Significant potential also exists for value-added treatment of aquaculture products. At present the Canadian industry is operating below potential and cannot meet international market demand.

EXAMPLE: JOB CREATION IN NEW BRUNSWICK SALMON FARMING

World food production over the next 15 to 20 years will change phenomenally from what we know today – and fisheries will be no exception – the global fish and seafood industry is undergoing a profound transformation that has significant consequences for Canadian fisheries policy. Global demand for fish and seafood is projected to grow steadily from under 100 million tons in 1990 to 120 million tons by the end of the decade. In recent years however, wild fisheries catches have peaked at around 100 million tons and have begun to decline. In fact, the majority of the world's 17 major fisheries are in serious decline. Global production of aquaculture products will help fill the projected supply/demand gap, and world production is expected to top 20 million tons by the year 2000.

The New Brunswick, Canada salmon farming industry began in 1978 on a pilot scale with the first successful attempt to over-winter Atlantic salmon in sea cages. By 1979 the first commercial aquaculture firm was established and produced 6.3 tons worth $45,000. The federal and provincial governments established a Cooperation Agreement on Fisheries Development to provide financial assistance to entrepreneurs in aquaculture to establish infrastructure. Twelve farms were established under this program and most remain in business today.

By 1993, New Brunswick had 46 firms operating 56 sites producing 11,000 tons of product valued at $90,000. Salmon farming is the major activity and employs over 460 full time employees, plus more than 1000 jobs in the related supplies and services sector. The industry has invested in 13 hatcheries, eight processing plants and seven feed mills – predominantly in coastal communities. This Canadian industry has now expanded hatchery and grow-out operations into Maine, USA, and the industry is looking to expand into other commercially important species such as halibut, haddock and striped bass.

Source: Stechey, Daniel, Shawn Connors and Robert H. Cook. (1994). "Aquaculture; A Model for Sustainable Economic Development in Canada." Fisheries and Oceans Canada, Ottawa, Canada.

EXAMPLE: PRAIRIE AQUACULTURE AT AGPRO

AgPro Grain Inc., a wholly owned subsidiary of the Saskatchewan Wheat Pool, has moved to diversify its operations by expanding into areas other than its traditional core business of grain handling. Aquaculture was one area chosen as a business that was economically viable, environmentally friendly and sociably acceptable. AgPro's initiative into aquaculture complements the Saskatchewan Wheat Pool's objective of sustainable rural development in the province.

AgPro Grain Inc. (AgPro) is currently 100% owner and operator of the Lake Diefenbaker Cage Culture operation. The fish farm is in its third year of operation, with 1994 sales expected in excess of one million dollars. The current fish farm has approximately 200,000 kg of fish in inventory, with a market value of one million dollars. This makes AgPro one of the largest trout producers in Canada.

The aquaculture business of AgPro Grain Inc. has resulted in the creation of 20 permanent full-time and part-time employment positions. The business has also created indirect jobs within Canada and the local community. Examples of aquaculture-generated employment in various sectors of industry including trucking, food processing, and manufacturing. Employment opportunities also exists for general labour, fish processing, waste management and aquaculture specialists. When possible, local residents have been employed as an incentive to stimulate economic opportunities in the immediate area.

The individuals hired by AgPro are also trained by AgPro. This is required as there are no skilled aquaculturists in Saskatchewan and very few skilled fish processors in the immediate area. Government assistance for this type of training would be invaluable as it is expensive.

Because Canada is a world leader in environmental protection, it is one of the most difficult countries to receive environmental approval to farm fish. Environmental approval from the government is the single largest barrier to entry into the aquaculture industry in Canada due to the sheer number of departments and agencies that have jurisdiction over Canada's water bodies. AgPro's initial application took three years before approval was finally obtained.

Source: Bielka, John. (1994). "Prairie Agriculture at AGPRO." AgPro Grain, Inc., Birsay, SK, Canada.

Aquaculture development has the potential to provide both much needed food stuffs and employment at a time when natural fish stocks are in serious decline.

BUILDING RETROFITS

Background

- The industry of environmental "retrofitting" of buildings is a new growth area in Canada. This industry reduces CO_2 emissions and conserves energy.
- In Canada, 90% of all homes will still exist by the year 2000. The construction of new houses will decline, and renovation and retrofitting will be a growth industry. There are between three to seven times as many more jobs per million dollars of investment in retrofitting than in new construction. Retrofitting is more labour intensive and uses less virgin building material.
- One of the key issues is the need to train both professionals and the general public regarding the benefits of renovation and ecological retrofitting.
- Job opportunities exist in home energy systems, building diagnostics, lighting retrofits, furnace replacement, mechanical upgrades, and retraining. Conservation programs put retrofitters back to work.
- New policies are required to ensure that retrofits and new construction address environmentally "friendly" considerations. This fledgling industry requires support in terms of credit access, training, certification, and public promotion/awareness.

CASE STUDY — BOYNE RIVER ECOLOGY CENTRE

The Coalition for a Green Economic Recovery includes individuals, businesses and organizations interested in developing and promoting appropriate renewal strategies based on environmentally sustainable partnership initiatives. In providing a clearing house function, the

Coalition is able to provide numerous examples of community pilot projects.

The Boyne River Ecology Centre is one such example which demonstrates what can be achieved with a fully integrated or holistic design process. The Centre integrates the design, construction, and operation of the building with its educational function and its surrounding environment. The Boyne is an autonomous building, relying on site-generated hydro, solar electricity, and solar and biomass heating. An indoor biological regeneration system or "living machine" turns human waste into pristine water and doubles as a fully integrated biology demonstration. The building is sheathed in glass for solar gain and to facilitate observation of the natural environment. The building, its infrastructure and its operation, demonstrate the fundamentals of environmental responsibility to visitors and provide a rich teaching tool and educational setting. The Boyne is an example of where we can and should be, not where we might think about being, in the future.

Source: Lowans, Ed. (1994). "Eco-Efficient Buildings." Coalition for a Green Economic Recovery, Orangeville, Ontario, Canada.

Through a holistic design process, retrofitting technologies can provide jobs, ensure sustainability and educate the public.

VALUE-ADDED INDUSTRIES

Background

Canadians generate roughly twice as much garbage per person as residents of France, Germany or Italy. Cheap and available energy and landfill space have made a materials intensive economy affordable. Growing awareness of the environmental costs of extracting, transporting, processing, packaging, distributing, consuming, and disposing of a wide range of products and materials have spurred interest in reduction, reuse, and recycling initiatives. Redesign, the strategy best able to reduce waste, is only starting to be seriously contemplated by public and private interests.

Reuse and recycling programs are both more labor intensive and resource conserving than traditional production and disposal practices. The relatively low-skilled labour required to perform most collection and sorting functions also provides an opportunity to integrate employment challenged individuals into the work place, replacing dependence on social assistance with earned wages.

Rapid growth in residential curbside recycling programs has dramatically increased the collection of secondary materials over the past several years. Processing capacity has not expanded as quickly in many areas, leading to market gluts and depressed prices for materials collected. General attributes of value added industries are:

- The growth of value-added natural resource and synthetic product industries has, in some cases, been exponential in recent years. Value-added processing is perhaps best known in the recycling industry where hundreds of new products can be developed from waste product.
- In the agricultural sector, the value-added applications of traditional agricultural products can be highly diverse. Sunflower seed shells for example can be burned in stoker type stoves to replace fossil fuels. One ton of shells can produce equivalent heat to 120 gallons of #2 fuel oil.
- The economic spinoffs of value-added activity can provide clean and efficient alternative fuel sources (e.g. ethanol, canola biofuels and lubricants, seed shells), environmentally friendly products; export products; and import substitutes.
- One of the disincentives to developing value-added products is the initial cost factor which is often two or three times the price of conventional fossil fuels, fertilizers, pesticides, herbicides, and virgin natural materials.
- Community-shared agriculture and processing is a growing phenomenon which places value on organic products and omit processing.
- In urban areas, recycling has provided commercially important examples of employment generation as well as new product development.

CASE STUDY — THE EDMONTON RECYCLING SOCIETY

The Edmonton Recycling Society (ERS), a community based non-profit organization, was awarded a contract in 1988, by the City of Edmonton, Canada, to collect, process, and market recyclable materials from the solid waste stream of the private residences in the north half of the City. The Society's mission is to "conserve creation and create employment".

The ERS has utilized local community resources in all aspects of its operation, including the design and manufacture of its equipment, the development of personnel policies, provision of training, and special support services for its employees, and the necessary financial backing to ensure viability.

Currently in its sixth year of operation, the Society has provided training in life skills and employment to more than 500 persons and steady, full time employment to a workforce of more than 70 persons. The program has exceeded its employment goals, its level of public participation, and its financial targets. During this time it has remitted over $500,000 of revenue to the City. It is estimated that the sales value of the materials manufactured from the recycled refuse is approximately seven to ten times that of the sales revenue for recyclables obtained by the society.

The economic impact of the ERS is approximately $4.5 million annually over and above its own operating costs. More than a dozen industries in the region are currently manufacturing products using recycled commodities as raw materials. Only the cleanest, highest quality material is marketable, locally and abroad. Hence the plant operates efficiently and produces clean, high quality recycled materials (e.g., cardboard, newsprint and magazines, milk cartons, mixed paper, telephone directories, five types of plastic, five varieties of metal including aerosol cans, glass, and refundable beverage containers of all kinds) which bring some of the highest prices on the continent.

In the past, jobs existed only because of unsustainable activities. The ERS believes that value-added processing, a focus on new and expanding niche markets, longer term investment horizons, a broader definition of corporate and worker responsibility, and a deep commitment to the development of self-sustaining communities are the way of the future.

Source: Guenter, Cornelius. (1994). "The Edmonton Recycling Society – An Experiment in Employment and Sustainable Development." Edmonton Recycling Society, Edmonton, Alberta, Canada.

By combining a community-based approach with environmentally responsible principles, the ERS has been able to develop a successful and sustainable business.

TRANSPORTATION

Background

- Transportation in the North American context has been described as a "monoculture" in which the personal automobile has squeezed out other modes of transportation at the expense of the environment. Current research focusses on the identification and development of alternatives to fossil fuels, solar powered electric vehicles, and conventional vehicle alternatives. Rail system renewal with high speed and frequent service is providing useful alternatives to traditional commuting by personal vehicle.

- Innovative research is being undertaken to look at the feasibility of "intelligent" vehicles or "smart cars" – e.g., vehicles which are less expensive, more efficient to operate and less damaging to the environment.
- A shift in orientation is occurring in terms of people-centred as opposed to car-centred planning. The principles focus on how to integrate transportation modes, e.g., walking, cycling, train, vehicles with residential and commercial needs.
- As a growth area, more and better jobs are being related to sustainable transportation, e.g., public transportation than to highway infrastructure. Public transport returns more net benefits to society. The present "rethinking" in urban design represents a shift towards a mix of transportation modes with corridors and feeder lanes. A new emphasis on cycling as a growth industry is characterized by "pedicabs", bicycle delivery, and police services. Tourism and recreational cycling is a growth area in many areas of Canada, notably Quebec and Nova Scotia. Cycling to work is now encouraged by many employers and companies routinely provide bicycle storage facilities to employees. The world of cycling has become "in vogue" and is recognized as a popular symbol of the individual's ability to modify lifestyle and behavior for the betterment of the environment.

ILLUSTRATION — SUSTAINABLE TRANSPORTATION

According to Sue Zielinski of Transport Options, automotive fuels account for the following: 17% of global carbon dioxide releases, two thirds as much as rainforest destruction. Motor vehicle air conditioners are the world's single largest sources of CFC leakage into the atmosphere. The USA consumes 40% of the world's gasoline. US reserves of oil will be depleted by 2020 and world resources by 2040. It is worth note that 47,000 Americans are killed each year on US roads, similar to the number that died in the Vietnam war. An additional 30,000 deaths each year are attributed to smog caused by motor vehicle emissions.

The Watershed Sentinel reports that the city of Eugene, Oregon, proposes to spend $157 million over the next seven years on four new highways. The environmental group Auto Relief did some research and found that the same amount could be used to purchase all of the following:

- Free bus transportation to the year 2000, including a $1 million per year administrative budget increase;
- A 10-fold increase in funding for ride-share programs;
- A free bicycle for every resident over the age of 11, complete with lock, helmet and rain gear;
- 10,000 free carrying racks for cars;
- 10,000 free bike trailers;
- 48 km of bike paths;
- Total elimination of the projected municipal debt by the year 2000;
- A $100 rebate to every citizen;
- A $7.1 million surplus.

Source: Zielinski, Sue. (1994). "Transporting Ourselves to Sustainable Economic Growth." Transport Options, Toronto, Ontario, Canada.

Money previously spent on the maintenance and development of highways can be used to develop alternatives which are less detrimental to the environment.

OCEAN RESOURCES AND COASTAL COMMUNITIES

Background

- The advent of the UN Law of the Sea (UNCLOS) and the extension of 200 mile exclusive economic zones (EEZs), gave coastal nations the world over many extra square kilometres of territory and new ocean wealth. One of the challenges to the marine nations is how to manage this new wealth in rational and sustainable ways.
- Opportunities for job creation in the marine sector include transportation, oil and gas exploration/extraction, sand and gravel extraction, harvesting of commercial and under-utilized species, aquaculture, biotechnology and pharmaceutical research.

- Three quarters of the world's population will reside in coastal communities by the year 2000. With the decline of commercial fish stocks, new forms of employment must be found which respect the integrity of the environment.
- Coastal waste management systems are inadequate. For example, 90% of Newfoundland has no sewage treatment system, and major urban centres such as Halifax pump raw sewage into the harbour. Sewage treatment is an imperative.
- New ecotourism livelihoods are being developed in coastal areas based on marine archaeology and historical sites.
- With the collapse of commercial fish stocks on the Atlantic coast, new forms of "value-added" processing and the harvesting of under utilized species offer communities new options for job creation.

CASE STUDY — TAKING STOCK OF COASTAL COMMUNITY STRENGTHS

In 1990, the Economic Recovery Commission of Newfoundland and Labrador initiated a project to examine the province's potential to develop new employment in sectors of the economy that present opportunities for growth outside the established resource industries. Eleven new growth areas were identified: manufacturing, innovative technology, information industries, export services, environmental industries, energy efficiency and alternative energy, aquaculture, adventure tourism, cultural industries, crafts, home-based and micro-industries.

The Commission has dedicated itself to:

(i) Reducing dependency on government;
(ii) Strengthening the private sector;
(iii) Diversifying the provincial economy.

Newfoundland presents a special case study in which, in the past, the provincial wealth was based almost exclusively on the export of raw and semi-processed materials: fish, wood pulp and paper, minerals and hydroelectricity. In the future, while these will continue to be important, the province must improve its competitiveness in the world export market through increased productivity. The Strategic Economic Plan for Newfoundland and Labrador has been guided by a policy of building upon strengths (rather than focusing on totally new industries that are alien to the province) and looking for new competitive advantages. With public/private sector partnerships and consultation, Newfoundland has undertaken an industry profile for each new growth industry. This strategic planning exercise has enabled the province to take stock of human and industry sector resources. The Commission is also addressing bureaucratic and government level barriers and is recommending new policy avenues to create a more positive climate for economic growth.

Source: House, Doug. (1994). "New Opportunities for Growth by Newfoundland Economic Recovery Commission." Chairperson, Economic Recovery Commission, St. John's, Newfoundland, Canada.

By taking stock of a region's resource base, in both physical and human capital terms, appropriate development strategies can be developed.

SUMMARY

In conclusion, there are a number of important considerations which affect the successful outcome of social development programs. The following summary draws conclusions from present practice and experimental initiatives, several of which have been described in this Chapter.

1. There must be a fundamental change in the culture and functioning of government. The appropriate role of government policies, programs, and agencies is to provide leadership and facilitation and to regular and co-manage environmental resource utilization.
2. Achieving sustainable development requires management strategies tailored to localized specific needs, not a top-down approach.
3. Better use of existing funds can finance needed investments. Accountability to those affected by spending decision is crucial to ensure the most productive use of scarce financial resources.
4. Improving the capacity of individuals, companies, and communities to act on and realize opportunities in the appropriate role of training programs, regulations, and government spending.
5. Cooperative co-management of natural resources is the best approach to promoting long-term stewardship. Combining public policy goals, sound science, and local experience and knowledge results is the wisest and most responsive use of resources.

6. Holistic, interdisciplinary approaches to problem solving result in the lowest cost solutions. Accomplishing multiple objectives simultaneously tends to lower costs, reduce environmental burdens, and satisfy the greatest number of people.
7. Decisions and development strategies that recognize and respect social and spiritual values will be widely accepted and endorsed. Ignoring these values will result in conflict.
8. Linear approaches to production and consumption violate every biological system we know. Protecting and restoring environmental integrity requires cyclical systems that eliminate waste.
9. Obtaining maximum value from human, financial and natural resources will provide the greatest gains for the lowest overall cost.
10. A new framework is needed to solicit the best ideas from public, private and community organizations. Effective alliances can serve the needs of all parties. They should be spear headed by independent groups who foster partnerships and promote synergies.

Some of the sustainable activities identified as significant potential employment generators in Canada and potentially for other parts of the world include:

Infrastructure Improvements

- Comprehensive building retrofits to upgrade energy, water, waste, and indoor air quality systems;
- Development and use of alternative transportation systems relying on walking, bicycles, mass transit, and vehicles powered by solar, biomass, hydrogen, and electric sources;
- Greatly expanded use of renewable energy and cogeneration technologies to provide electricity;
- Expanded tree planting and green space development in urban areas;
- Sewage and wastewater treatment systems that incorporate and mimic biological processes.

Knowledge Access and Use

- Broadening, extending, and enriching the information highway in locations across Canada and internationally;
- Widespread computer literacy training;

- Distance learning centers using interactive data bases and information and video networks to teach skills tailored to local communities and specific individuals;
- Expanded mapping of existing land uses delineating areas of ecological, cultural, economic, and religious significance;
- Improved inventories of the ocean and forest floor, wetlands, and other environments to better inform resource management decisions and provide a benchmark against which to measure change;
- Passing on and incorporating traditional knowledge;
- Improved management and harvesting of forestry, agriculture, and fishery resources;
- Adding value to commodities – agricultural, forestry, fishery, mineral and chemical;
- Increased reliance on aquaculture to satisfy a growing demand for fish and seafood products;
- Assessing the economic potential of underutilized species – ecosystem protection and restoration.

Development, Commercialization and Use of Environmental Technologies

- To reduce the consumption of energy and materials;
- To collect, sort, and process recyclable materials so they are used again in products of value;
- To clean up, reduce, and eliminate waste streams;
- To restore degraded environments;
- To comply with international environmental agreements covering ozone depletion, climate protection, and biodiversity preservation.

Service Industries

- Provision of quality child and elder care;
- Tourism development based on sites of environmental, cultural, and historical interest and significance;
- Lawn and garden care that incorporate ecological principles and biological controls;

- Increased reliance on conflict avoidance and mediation to resolve resource-based disputes;
- Health maintenance and disease prevention;
- Improved marketing of environmentally and socially responsible goods, services, and technologies;
- Environmental research, assessment and management services.

"There are numerous barriers to implementing the policies needed for sustainability. However, contrary to much popular opinion, progress is impeded more by cultural and sociopolitical factors than by technological impediments. Serious action is blocked by popular ignorance and scientific uncertainty wedded to cultural conservatism, the power of vested interests and political inertia. This general paralysis is reinforced by popular fallacies that sustainability can be achieved through gains in technological efficiency alone and that knowledge and service-based economies are inherently more ecologically benign than primary and manufacturing economies." (Rees, 1994).

Despite such sobering thoughts, bordering on what some might call pessimism, the examples presented here demonstrate the "art of the possible" through innovative example and creative optimism. The pathway to implementation will require significant commitment and dedication of resources, particularly to define problems more precisely, to mobilize energies, and to promote the sustainable development vision. It is at this juncture that the World Summit for Social Development can set the stage for a consultative mechanism to advance the principles of sustainable employment and social progress.

CHAPTER 5 SOCIAL SUMMIT OUTPUTS

INTRODUCTION

As we have seen in Chapter 4, there exist encouraging examples of communities, private sector and government joint efforts which are attempting to address social development problems and job creation through collective action and cooperative means. We hope the examples presented in Chapter 4 are able to "break new ground" in terms of identifying "best practise" in community-based employment creation. IISD has undertaken to codify best practice and to identify criteria for employment creation, while remaining faithful to concerns of economic, environmental and social impact.

The WSSD promotes an underlying theme of fostering a renaissance of spiritual and moral values. The overall goal suggests that each society should be able to integrate its members in a harmonious manner. At the same time, the WSSD acknowledges that the moral fibre of contemporary societies is being eroded. While the themes of "rebirth", "renewal" and "renaissance" are valid, practical means are required to convert these altruistic ideals into working constructs for advancing the quality of human life.

Despite the remarkable advances of humanity in scientific, technical and biotechnical areas, very few if any, of the world's communities have succeeded in eradicating poverty, discrimination, social insecurity. In essence, despite technological advances, there has been little improvement in these areas over the last three thousand years.

While the WSSD affirms that "social progress is built upon respect for the dignity of each person, development of the material and spiritual well-being of each community and the solidarity which must bind groups and nations" there are few working examples of these principles in action.

RECOMMENDATIONS TO WSSD DELEGATES AND SECRETARIAT

Primed with this tripartite social and environmental philosophy, we suggest the following recommendations for advancing the cause of global social progress and for improving existing delivery systems and international machineries. These recommendations are submitted for consideration and use by the Social Summit delegates and secretariat. We hope they will be useful in enhancing the deliberations of the Preparatory process, and influencing the outcome of the Summit.

After 45 years of multilateral, donor and international financial aid, notably from the UN family of agencies, the Brettons Woods Institutions and regional banks, a comprehensive and integrated global review of social and economic development policy and operations is overdue.

In tandem with paragraph 205 of the Draft Programme of Action Secretariat Doc. A/CONF.166/PC/L.13, the functions of monitoring social development and social progress should also be entrusted to a variety of international and national NGOs to ensure arms length perspectives, greater transparency, and accountability. These outcomes should be considered in an overall review of the international development system.

This review should be undertaken both within and outside of the UN system, primarily by a coalition of international NGOs, and should address operational as well as developmental lessons learned. The study mandate should include, *inter alia*, in depth assessment of the day to day operations, the cost effectiveness, efficiency levels, accountability issues, and quality of product and service as delivered by the international community and international civil service. The achievements, outcomes and lessons of this global study should be applied to the implementation of the WSSD Programme of Action. Specific comment should be made with respect to enhancing conventional aid delivery systems, operational and technical infrastructure, and accountability to the world's tax payers.

The provision of 0.7% of national official development assistance (ODA) targets (even if realised) will be insufficient to fund the Programme of Action as articulated in the draft documents (refer to item 214b). An international conference of ODA donor agencies and counterparts is required to undertake cost-benefit forecasting and commitments with respect to the real and actual costs for the proposed programme of action.

Throughout the Draft Programme of Action Secretariat Doc. A/CONF.166/PC/L.13, reference is made to the concept of "containing the spread of disease", including HIV/AIDs. This containment concept is better replaced with the concept of "disease eradication and humane treatment for all afflicted peoples". The notion of disease requires no passport, and containment efforts in poor countries will only exacerbate the situation and serve to greater isolate developing regions.

While the UN may provide a genuine "social pillar" for international cooperation (refer section 207), certain functions of social progress may flourish better through a parallel non-governmental body.

Furthermore, discrete functions will require arms length processes, notably monitoring and evaluation. The UN family of agencies should concentrate their respective roles in terms of existing strength areas and focus upon identifying new ways of enabling and facilitating national governments, NGOs and private sector partnerships to assume leadership roles.

Following the conclusion of the Summit, the Programme of Action will require extensive funding and political commitment from all nations. The process of social and financial mobilization will, at the same time, require a fundamental shift from the economics of industrialization to restorative economics in which consideration for the planet's ecosystem is central to all social and economic decision making. The private sector will play a major role in the reconstruction process. Efforts are required to reach out to private sector leadership and to involve business in the process of social development.

A critical feature of the Programme of Action which is still required is an appropriate institutional arrangement for implementation of the action plan. In this context, we recommend the creation of an international NGO body to complement the work of the UN family of agencies and to undertake integrated and comprehensive social progress enabling functions, with a special focus on integrating community and private sector interests.

It is proposed that this International Council on Social Progress undertake to provide service with respect to the coordination and implementation of certain essential features of the social progress program. The Advisory Board of the Council will be made of representatives from the private sector, government and community organizations. The Council international headquarters can be situated in Denmark, the host country for the WSSD; in Chile, the lead country in the development of the idea on the Summit or in a region or country, which has experienced enormous social disruption and change due to the decline of commercial stocks of fish, loss of traditional resource industry, and high unemployment. It is crucially important that the Council be hosted by a country with extensive experience in the multi-stakeholder process.

This Council will address problems and issues from an integrated sustainable development perspective. Key responsibilities will address current service delivery gaps, namely the development of tripartite community consultation models; monitoring; evaluation; business liaison; roundtable brokerage services; sustainable employment and development research; employment and debt reduction studies; and provision of assistance to NGOs, industry and national level machineries vis a vis the adoption of new ideas and practises of "restorative economics", community consultation and co-management of natural resources, and community governance. The Council will not serve as a project executing agency.

Additional potential areas of responsibility for the Council may include, but are not necessarily limited to, the following:

- Identification of simpler delivery mechanisms for managing change;
- Definition of problems;
- Mobilization of community, NGO, business, and government energies;
- Provision of backstopping policy services to national government and NGO machineries and implementing bodies;
- Promotion of policy coherence and research on social progress, reduction and elimination of poverty, debt reduction, restructuring of the UN agency and international financial institutions;
- Monitoring of national and global strategies for social progress;
- Provision of consultative support for overall assistance needs, and facilitation of the needs assessment process;
- Coordination of a clearing house function, data bases on employment and sustainable livelihoods, baseline data collection;
- Provision of a Social Progress Bulletin or newsletter along the lines of the IISD Earth Negotiations Bulletin. The unique feature and service of this bulletin will be the opportunity for marginalized groups to provide feedback through electronic-mail linkages, community conferencing and roundtable processes;
- Facilitation and negotiation of international television coverage of future summits, regional meetings, national fora to ensure greater transparency of the process and opportunities for feedback;
- Foster international exchanges and specialized training, practica, etc.;
- Encourage and facilitate greater donor collaboration, particularly with respect to promoting/harmonizing

common proposal applications and evaluation guidelines, sharing of evaluation results and "lessons learned", project results and outcomes.

It is envisaged that the Council will have an initial five year mandate and funding base. By the end of this period the Council should be self-financing through contributions from government foundations and the private sector. As a membership based organization, the Council will have national council counterparts and a provision for institutional and individual members. In certain respects the body will resemble the structure of the Earth Council which was formed following the conclusion of the UNCED. The International Council on Social Progress however will be accountable to national councils and its membership at large.

The outputs of the WSSD should therefore be expanded to include a revised Programme of Action, the Copenhagen Declaration, the UN Charter of Social Progress, an international public works program and youth voluntary service and the creation of a new non-governmental International Council on Social Progress. It is recommended that an international conference be convened prior to the end of 1995 to create the policy and working framework for such a Council. Representatives from industry, community groups, national governments, NGOs and the UN family of agencies should be invited. The conference should be funded from present and future commitments to the WSSD and the Programme of Action.

CONCLUSION

We conclude with the hope that the ideas contained in the vision, the transitions, the case studies and the suggestions for the outcome of the Summit might indeed be reflected in the declaration and action plan which are finally adopted. Cognisant of the nature of the negotiations and the process of consensus building which finally determines the agreements adopted in such Summits, our expectations in this regard are tempered. Our effort would nevertheless have been amply rewarded if these ideas were to stimulate thought and discussion (inside or outside of the Summit process) which ultimately leads to meaningful and concrete action towards sustainable livelihoods, poverty eradication and social integration.

At a minimum, it was our intention to demonstrate practical opportunities for hope and optimism.

BIBLIOGRAPHY

Barnet, Richard. "The End of Jobs" in *Third World Resurgence*, Vol 44, April 1994.

Brascoupe, Simon. "Sustainable Cultural Development: Sustainable Development in the Past and Future of Aboriginal Employment in Canada." Apikan Indigenous Network, Ottawa, 1994.

Chambers, Robert. *Rural Development: Putting the Last First.* Longman, Scientific and Technical, Essex, 1983.

Chossudovsky, Michel; "Brazil's Debt Saga" in *Third World Resurgence*, Vol. 44, April 1994.

Clairmont, Fredric. "The G-7 and the Spectre of Job Destruction" in *Third World Resurgence*, Vol. 44, April 1994.

Hawken, Paul. *The Ecology of Commerce.* Harper Business Press, N.Y. 1993.

Human Development Report. Oxford University Press, Cary N.C 1993.

Human Development Report. Oxford University Press, Cary N.C., 1994.

Idris, S. M. Mohamed, Chief Editor. "The Worst Global Unemployment Crisis Since the 1930s – ILO" in *Third World Resurgence*, Vol. 44, April 1994.

Khor, Martin. "Worldwide Unemployment Will Reach Crisis Proportions Says Social Expert." (quotation of Jeremy Rifkin "The End of Work"), *Third World Resurgence*, Vol. 44, Apr 1994.

Korten, David. "Development Heresy and the Ecologic Revolution." The People-Centred Development Forum, Manil January 2, 1992.

Neave, David and John Girt. "Linking the Solitudes of Wildlif Habitat, Landscapes, and Economic Development to Creat Sustainable Employment Opportunities." International Institut for Sustainable Development, Winnipeg, 1994.

Our Planet. United Nations Development Programme, Nairobi, Vol. 6, No. 2, 1994.

Rees, William E. "Sustainability, Growth, and Employment: Toward an Ecologically Stable, Economically Secure, and Socially Satisfying Future." International Institute for Sustainable Development, Winnipeg, 1994.

Said, Edward W. *Culture and Imperialism.* Alfred A. Knopf, N.Y., 1993.

Sen, Amartya. "Poverty and Famines: An Essay on Entitlement and Deprivation." Clarendon Press; Oxford, 1981.

Singh, Naresh and Richard Strickland. "Sustainability, Poverty and Policy Adjustment – from legacy to vision." International Institute for Sustainable Development, Winnipeg, 1993.

Singh, Naresh and Vangile Titi. "Adaptive Strategies of the Poor in Arid and Semi-arid Lands: In Search of Sustainable Livelihoods." International Institute for Sustainable Development, Winnipeg, 1994.

Stackhouse, John. (1994). "Developed but Still Destitute" in *Globe and Mail,* Saturday, July 23 Section D, p.1.

Third World Resurgence. No. 44, April 1994.

Watkins, Kevin. "World Bank, IMF Responsible for African Misery" in *Third World Resurgence,* No. 44, April 1994.

World Commission on Environment and Development. *Food 2000: global policies for sustainable agriculture.* Zed Books, London, 1987.